Silicon Carbide Micro Electromechanical Systems for Harsh Environments

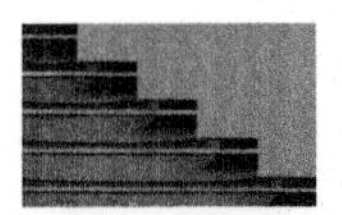

Editor
Rebecca Cheung
University of Edinburgh, UK

Silicon Carbide Micro Electromechanical Systems for Harsh Environments

Imperial College Press
ICP

Published by

Imperial College Press
57 Shelton Street
Covent Garden
London WC2H 9HE

Distributed by

World Scientific Publishing Co. Pte. Ltd.
5 Toh Tuck Link, Singapore 596224
USA office: 27 Warren Street, Suite 401-402, Hackensack, NJ 07601
UK office: 57 Shelton Street, Covent Garden, London WC2H 9HE

British Library Cataloguing-in-Publication Data
A catalogue record for this book is available from the British Library.

SILICON CARBIDE MICROELECTROMECHANICAL SYSTEMS FOR HARSH ENVIRONMENTS

ISBN-13 978-1-86094-624-0
ISBN-10 1-86094-624-0

Editor: Tjan Kwang Wei

Printed in Singapore

PREFACE

First of all, I would like to thank all the expert contributors to this book, without whom this project would not have been possible. My deepest acknowledgments also go to those at the Scottish Microelectronics Centre who in one way or another have contributed to the science and technology described in chapter one of this book. It has been the aim that a manuscript on silicon carbide (SiC) microelectromechanical systems (MEMS) be edited so that up-to-date information can be condensed into book form, easily accessible by the academic community as well as commercial companies. The present book contains high–quality information concerning SiC MEMS for harsh environments summarised and distilled for students, academics and researchers engaging in SiC MEMS to use and, I hope, serves as a valuable contribution to the MEMS community.

Microelectromechanical systems are essentially mechanical devices/sensors at the micro–scale. Applications of MEMS are wide ranging including for instance, miniaturised sensors for acceleration and pressure, wind sensors that mimic cricket hairs and microfluidic pumps for biomedicine. The main advantage of SiC as compared to Si is naturally the mechanical and chemical stability of the material. Once the material properties favours certain types of applications such as high temperature and harsh environments, there creates an impetus to advance the science and engineering in order to progress towards the final product. It is the science and technology that this book is concerned with, from the creation of the SiC material to its formation into the final microelectromechanical system.

The following five chapters combine to give an excellent review of the state-of-the-art technology and processes for the micromachining of SiC, growth of SiC, contacts to SiC, and etching of SiC, with the final chapter focussing on the applications of SiC MEMS.

Rebecca Cheung
Edinburgh, 2005

CONTENTS

CHAPTER 1

INTRODUCTION TO SILICON CARBIDE (SIC) MICROELECTROMECHANICAL SYSTEMS (MEMS)

Rebecca Cheung

School of Engineering and Electronics
King's Buildings
University of Edinburgh
Edinburgh, EH9 3JL, Scotland, UK
E-mail: r.cheung@ed.ac.uk

This chapter serves as a brief introduction to the basic properties of silicon carbide (SiC) and the advantages of using SiC over other semiconductor materials for microelectromechanical systems (MEMS). Given the excellent and extensive review chapters that follow this one, I have confined this chapter to recent research performed at the University of Edinburgh in the area of SiC microelectromechanical systems (MEMS). Some of the processes involved in the fabrication of microelectromechanical systems in SiC are discussed, together with the problems to be overcome in order for SiC's potential as a MEMS material be exploited in applications for harsh environments.

1. Introduction

The total MEMS market worldwide already exceeds $10 billion, up from $100 million only five years ago.[1] Addressable markets include automotive, medical, telecommunications, industrial, transportation and environmental while consumer products include household appliances and toys. Currently, the most successful MEMS sensors are made in silicon. Examples include sensors that trigger the deployment of automotive airbags as well as ink jet nozzles. On the other hand, commercial MEMS in SiC is still in its infancy and occupies a niche market. However, the future SiC MEMS market could become substantial, contributing a significant percentage of the total market for MEMS. Because of the unique material properties of SiC including wide bandgap, mechanical strength, high thermal conductivity, high melting point and inertness to exposure in corrosive environments, devices manufactured in SiC are more robust. Such SiC MEMS can operate at higher temperatures and in harsh environments compared to their silicon counterparts.[2,3] Potential new markets where SiC sensors could make a large impact include, for example:

(1) Radio frequency (rf) MEMS area for rf and millimetre wave applications in military, commercial wireless communication, navigation and sensor systems, where devices including micro-switches, tunable capacitors, micro-machined inductors, micro-machined antennas, micro-transmission lines and resonators made in SiC present potential improvements in operating frequency, power handling capability and reliability compared to devices made in silicon;

(2) Pressure sensors for use in the oil industry where currently, the procedure for oil drilling is modified in order to prevent sensors from physical damage due to the high vibrational environment;

(3) Accelerometers in aeroplane engines and motors in harsh environments for detecting acceleration, hence potentially providing better safety control;

(4) Optical MEMS (MOEMS) applied to general industrial applications for control of light, sensing and in manufacturing technologies.

RF MEMS is expected to be the third major player in the MEMS market, estimated to exceed $1 billion by 2007, while revenues for simple MEMS devices such as filters and inductors are predicted at a modest $200 million by 2007.[4] Long term reliability of components is believed to be the second most important issue after price. Therefore, SiC MEMS has extremely strong prospects as a key platform process. Similarly, optical MEMS for sensor systems is predicted to grow to $347 million in 2007 and $100 million for positioning and alignment systems.[5] The advantages of SiC for MOEMS is yet to be explored, but high stability is certainly one potentially exploitable parameter.

However, thus far, the difficulty in the growth and processing of the material has meant that progress in the use of SiC for MEMS applications has been slow. Nevertheless, in the past decade, tremendous efforts have been put into the growth and processing aspects of SiC and as a result, the application of SiC as a MEMS based material is beginning to appear attractive. The remaining chapters of this book combine to give an excellent review of the state-of-the-art technology and processes for the growth of SiC, contacts to SiC and etching of SiC, with the final chapter focussing on the applications of SiC MEMS.

2. SiC Material Properties

SiC exhibits a one-dimensional polymorphism called polytypism. All polytypes of SiC have an identical planar arrangement of Si and C atoms, which are distinguished by differences in the stacking sequence of the identical planes. Disorder in the stacking periodicity of similar planes results in a material that has numerous crystal structures (polytypes), all with the same atomic composition. The magnitude of the disorder is such that more than 250 SiC polytypes are identified to date.[6] Despite the large number of polytypes, only three crystalline structures exist: cubic, hexagonal and rhombohedral. The origin of the polytypism can be visualized as follows. In Figure 1, the solid circles represent spheres closely packed in a plane; call this "plane 1". To place another such set of spheres on top of plane 1 as closely as possible, one would place each sphere in the hole between any three neighbouring spheres in plane 1 (dotted circles, plane 2). But there is another way of accomplishing this: the dashed circles in plane 3. The order of stacking of the planes determines the types of close-packed structures and their symmetry properties. According to conventional nomenclature, a SiC polytype is represented by the number of Si-C double layers in the unit cell, the appending letter C, H, or R indicating a cubic, hexagonal or rhombohedral symmetry. For example, the 6H hexagonal lattice has six such layers in the primitive cell with the following succession of the above planes: 1,2,3,1,3,2,1,2,3,1,3,2; the 3C lattice is built up as 1,2,3,1,2,3; 2H-SiC corresponds to 1,2,1,2; and 4H-SiC corresponds to 1,2,1,3,1,2,1,3.

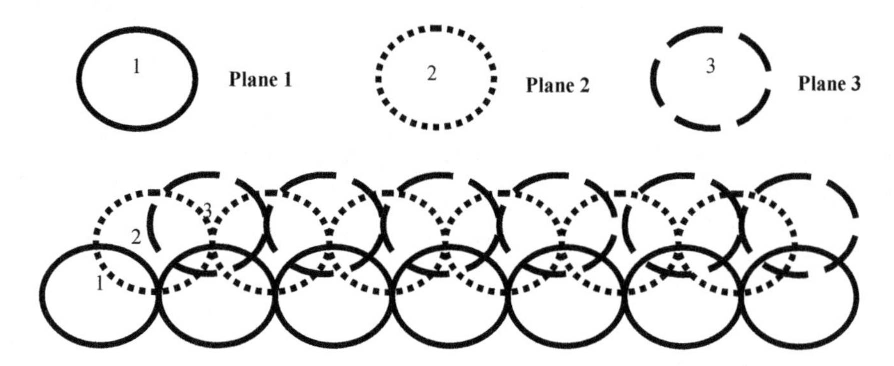

Figure 1. Schematic diagram of atomic arrangements in the different SiC polytypes (see text).

Although all SiC polytypes have the same atomic composition, the electrical properties differ. For instance, the bandgap for SiC ranges from 2.3eV for 3C-SiC to 3.4eV for 4H-SiC. Despite having the smallest bandgap, 3C-SiC has the highest electron mobility ($1000 cm^2/Vs$) and saturation drift velocity ($10^7 cm/s$), due in part to its cubic crystalline symmetry.

SiC has always been noted for its excellent mechanical properties, specifically, hardness and wear resistance. In terms of hardness, SiC has a Mohs hardness of 9, which compares favourably with values for other hard materials such as diamond (ten) and topaz (eight). In terms of wear resistance, SiC has a value of 9.15, as compared with 10.00 for diamond and 9.00 for Al_2O_2. SiC is not attacked by most acids and can only be etched by alkaline hydroxide bases (i.e. KOH) at molten temperatures (> 600°C). SiC does not melt, but sublimes at about 1800°C. The surface of SiC can be passivated by the formation of a thermal SiO_2 layer, even though the oxidation rate is very slow when compared with Si. The above properties are not generally polytype dependent. A comparison of the fundamental material properties of 3C-SiC, Si and 6H-SiC can be found on p.182 of this book and demonstrates the large potential of SiC MEMS as compared to silicon MEMS when applied in harsh environments.

3. Making a Microelectromechanical (MEM) Device

In order to make a microelectromechanical device, many aspects need to be considered including the growth of the required layers, design, processing, packaging and testing. SiC exists in different crystalline states, namely, single crystal, polycrystalline and amorphous. Different degrees of crystallinity can be grown on various substrates and as a result, a large combination of multi-layers containing SiC films can be possible and the processing for the final device of system depends on the layer design and the application. The chapters that follow provide an excellent overview of the constraints and possibilities in SiC film growth and processing for MEMS applications. In the following section, the two commonly employed processes including bulk and surface micromachining for MEMS applications are discussed.

3.1. Micromachining processes

3.1.1. Bulk micromachining

Conventional silicon bulk micromachining can be used for single-crystal, poly and amorphous SiC. For single-crystal SiC, the SiC must be grown directly on silicon. In this case, both front and back-side micromachining are possible as shown in Figure 2. Due to the high etch resistance of SiC, most commonly used anisotropic wet etchants can be used to remove the bulk silicon.

To improve reliability and control,[7-11] the group at Edinburgh has developed an all dry etch process for the bulk micromachining of 3C-SiC resonators on silicon.[12] A one-step inductively coupled plasma etch process using SF_6/O_2 gas mixture has been developed to fabricate straight resonators. The SiC resonators have been made first before the release of the cantilevers and bridges, performed by etching the silicon isotropically. The cantilevers and bridges have resonant frequencies between 120kHz and 5MHz depending on the device geometry; see Figures 3 and 4.

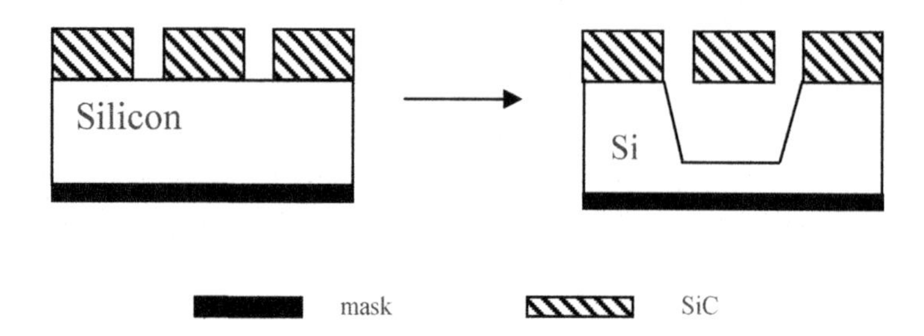

Figure 2(a). Bulk micromachining – release of SiC film via etching of silicon from the front of the wafer.

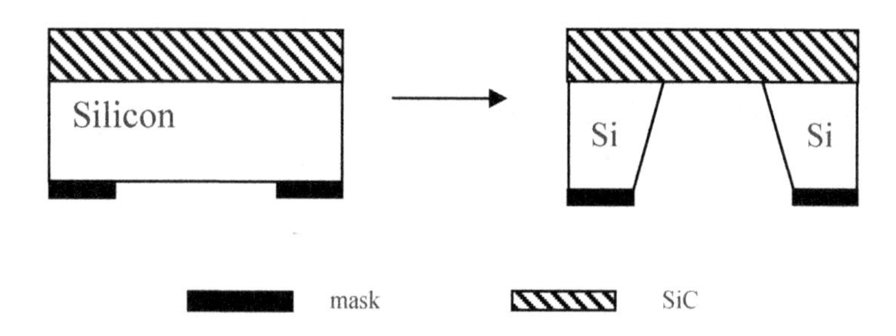

Figure 2(b). Bulk micromachining – release of SiC film via etching of silicon from the back of the wafer.

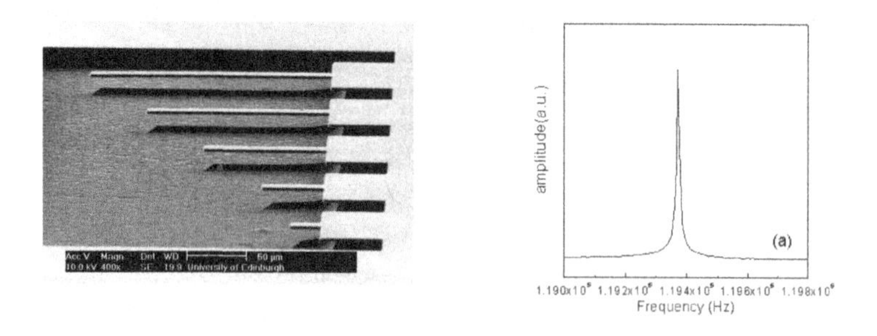

Figure 3. Array of SiC cantilever beams with lengths 25, 50, 100, 150, 200 μm, released from silicon using one-step dry etch process and the corresponding resonance response for the 200 μm cantilever beam.

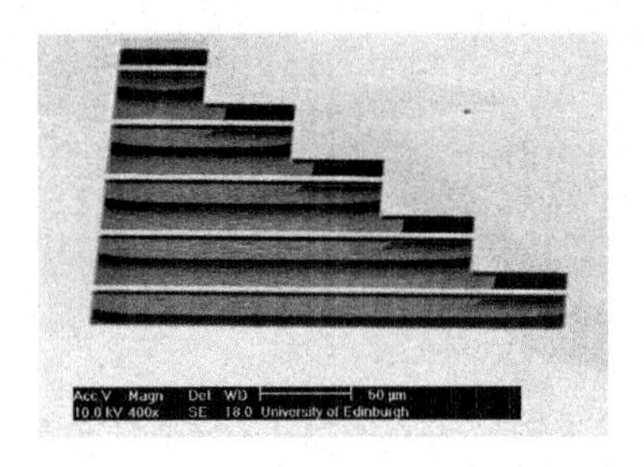

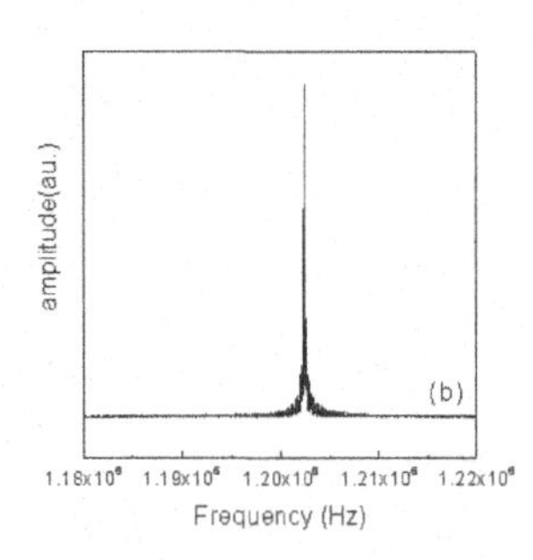

Figure 4. Array of SiC bridges with lengths 50, 100, 150, 200, 250 μm, released from silicon using one-step dry etch process and the corresponding resonance response for the 200 μm bridge.

3.1.2. Surface micromachining

The possibility to grow and process SiC films in multi-layer structures allows complex MEMS to be designed and processed and open doors to many applications. Similar wet or dry release processes employed for bulk micromachining can be used for surface micromachining also. For example, poly-SiC grown on a poly-Si layer or deposited on oxide layers, can be used as the mechanical layer while the poly-Si or oxide layer is used as the sacrificial layer, illustrated schematically in Figure 5. When poly-Si is used as the sacrificial layer, KOH, TMAH or our developed dry etch recipe can be used to release the SiC resonators, while the oxide is used to protect the underlying silicon during the sacrificial etch,[13] see Figure 6. An example of our fabricated resonator, 200 μm long, can be shown to be actuated electrostatically with a fundamental resonant frequency of 66.65kHz and an amplitude dependence on the applied V_{dc} and V_{ac} (Figure 7).[13] The processing of capacitively driven resonators and piezoresistive strain gauges on similar multi-layers[13-18] form the basis for more complex MEMS including accelerometers[19,20] and pressure sensors.[21-25] Further, for the first time, our 3C–SiC cantilever resonators have been shown to resonate upon electrothermal actuation.[26]

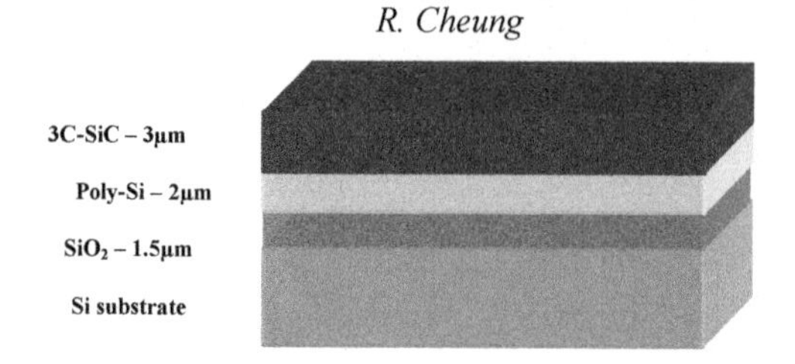

Figure 5. Multi-layer SiC/poly-Si/SiO_2/Si material structure.

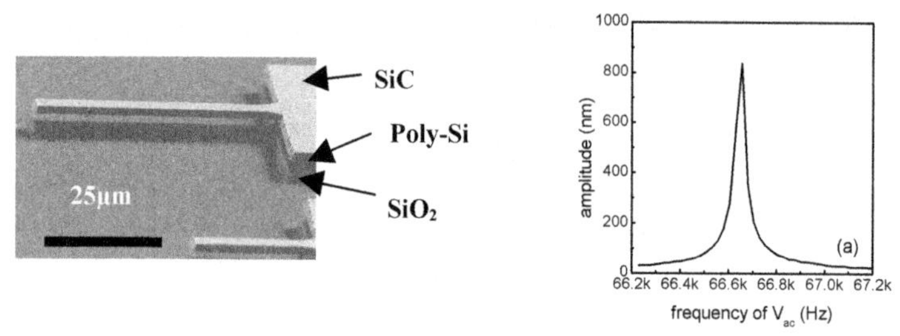

Figure 6. Cantilever beam fabricated in the multi-layer material structure and its corresponding fundamental resonant frequency.

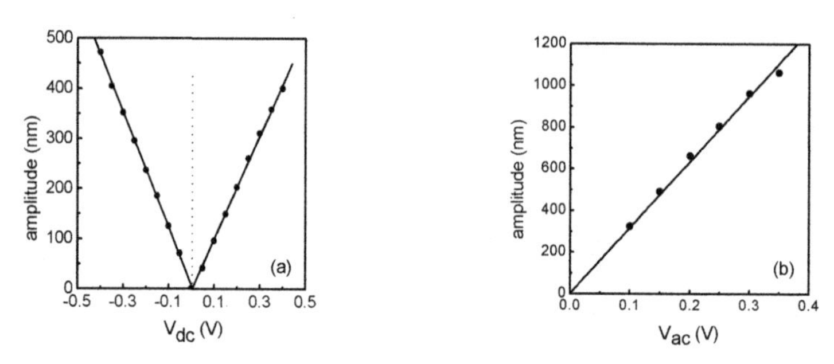

Figure 7. Amplitude Z for a 200 μm long cantilever as a function of: (a) V_{dc} (V_{ac}=0.3V), and (b) V_{ac} (V_{dc}=0.2V). The frequency of the applied a.c. signal was 66.65 kHz. The solid lines are linear fits to the data points.

4. Surface Modification

It has been reported recently that the surface microstructure of the MEM device can affect its response in particular the quality factor.[27] We have studied the surface modification at the microscopic scale after inductively coupled plasma etching of 4H-SiC in SF_6/O_2 using x-ray photoelectron spectroscopy.[28,29] Both C 1s and F 1s spectra from the etched SiC under various etch conditions have been analyzed and studied. Our findings show the existence of both covalent and semi-ionic C-F bonds on the etched SiC surfaces, probably due to the existence of reactive F ions in the plasma. The intensities of the components of C-F groups in the C 1s spectra have been seen to decrease with the increase of O_2 in the gas mixtures, see Figure 8(a). The higher concentration of O_2 in the plasma would serve to remove C atoms thus leaving less C atoms available to react with F, causing the C-F groups in the C 1s spectra to become weaker with the increase of O_2 concentration in the SF_6/O_2 gas mixture.

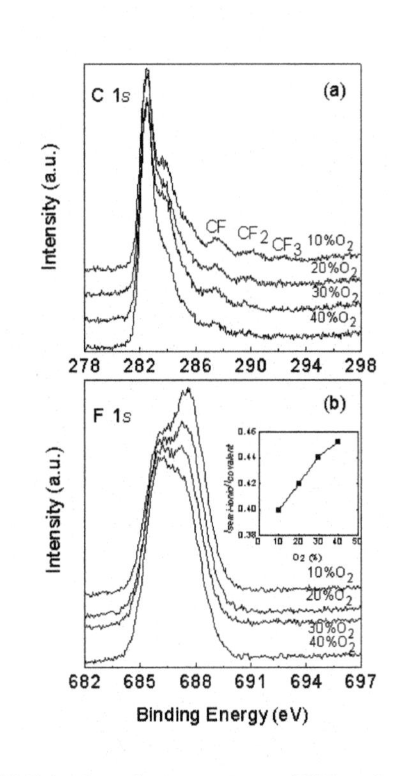

Figure 8(a). C 1s and (b) F 1s photoelectron spectra of SiC surface etched at different O_2 concentration in SF_6/O_2 gas mixture. The inset shows $I_{semi\text{-}ionic}/I_{covalent}$ ratio in the F 1s spectra as a function of O_2 concentration.

Figure 8(b) shows the F 1s spectra from the SiC etched at different O_2 concentrations in the SF_6/O_2 gas mixture. It can be seen that with the increase of O_2%, the dominant F 1s is changed gradually from covalent to semi-ionic C-F bonds. Shown in the inset is the increase in the relative $I_{semi\text{-}ionic}/I_{covalent}$ ratio with O_2% increase. In addition, it has been found that, in order to achieve higher etch rate of SiC, optimum O_2% in the SF_6/O_2 gas mixture and flow rate have to be applied during the dry etching of SiC. Furthermore, increasing chuck power and decreasing work pressure in the ICP system can also promote the etch processes. Figure 9 shows SiC etch rate and the F/Si ratio as a function of dc bias and chuck power, where larger etch rates are observed at higher dc bias and chuck powers. With the increase of chuck power, hence dc bias, more F incorporation is observed.

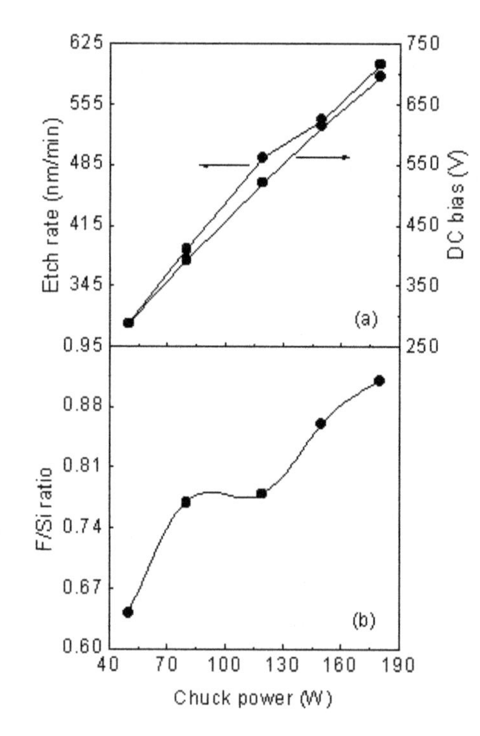

Figure 9(a). Etch rate and dc bias and (b) F/Si ratio vs applied chuck power. (SF_6 flow rate = 6-sccm, O_2 flow rate = 15sccm and pressure = 5mT).

From Figure 10(a), it can be seen that the intensities of the C-F bonds increase with the applied chuck power, particularly in the CF_2 bond. In addition, the CF_3 bond is observed only in the sample etched at 180W chuck power, while it is not obvious in the other two samples etched at lower chuck powers. These observations suggest that with the increase of the incorporated F concentration, the formation of C-F bonds is in the order of CF, CF_2, and CF_3, which is in agreement with the study of the growth of fluorinated carbon films.

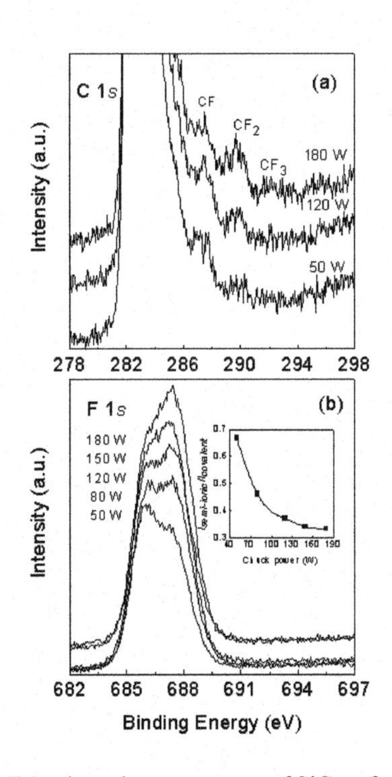

Figure 10(a). C 1s and (b) F 1s photoelectron spectra of SiC surface etched at different chuck powers. The inset shows $I_{semi\text{-}ionic}/I_{covalent}$ ratio in the F 1s spectra as a function of chuck power.

The F 1s spectra in Figure 10(b) show clearly the decrease of semi-ionic to covalent C-F bonds with an increase of the applied chuck power and etch rate, corresponding to the increase of the concentration of the incorporated F, as shown in Figure 9(b). Because of the difference in the electronic behaviour between semi-ionic and covalent C-F bonds,[30] our observation suggests that, SiC surfaces processed under lower etch rate conditions can become more conductive compared to those surfaces processed under higher etch rate conditions.

During the studies of F 1s spectra as a function of chuck power, pressure and SF_6 flow rates, it has been found that, in most cases, both the relative F concentration and the $I_{semi\text{-}ionic}/I_{covalent}$ ratio decreases as etch rate increases. Such microscopic surface modification on the subsurface could affect the MEM device performance such as the quality factor.[27] Moreover, the nature and quantities of the covalent and semi-ionic C-F bonds on the SiC etched surfaces, coupled with the complex etch mechanism of SiC,[31] can affect the performance of SiC electronic devices, as demonstrated in our recent study on the electrical behaviour of 4H-SiC Schottky diodes after inductively coupled plasma etching.[32-33] The effect of process–induced defects on the Schottky contacts is also discussed in Chapter 3.

5. Frequency Tuning of the SiC MEMS

More recently, our group demonstrated the resonant frequency tuning capability of SiC MEM resonators using focussed ion beam (FIB) deposited platinum (Pt),[34] see Figures 11 and 12. Platinum of surface area 13 x 5 μm^2 and thicknesses ranging from 0.3 to 3.1 μm has been deposited at room temperature on the cantilever and bridge resonators. The resonant frequency of the SiC cantilevers (Figure 11) and bridges (Figure 12) can be adjusted up to 12% by either adding platinum to or removing platinum from the resonators providing flexibility in tuning the resonant frequency when necessary.

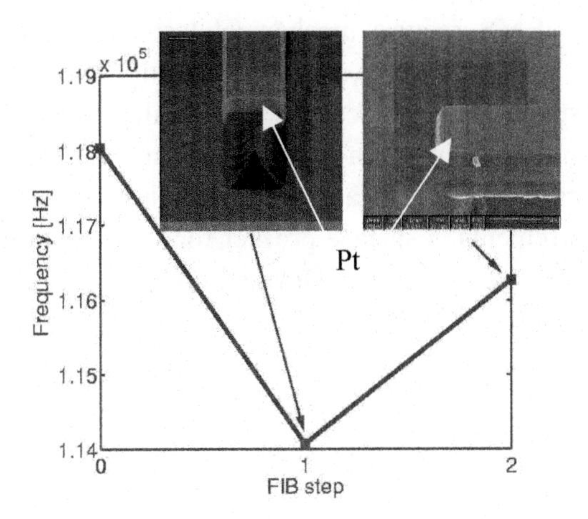

Figure 11. Change in frequency as a result of SiC cantilever (15 μm wide) tuning. FIB step 1: platinum deposition. FIB step 2: platinum milling.

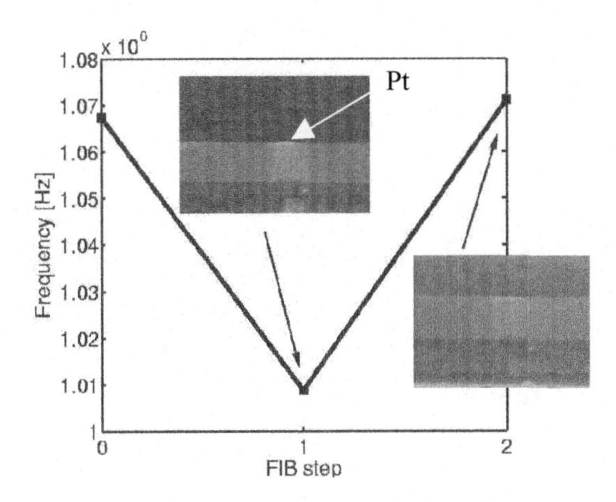

Figure 12. Change in frequency as a result of SiC bridge (15 μm wide) tuning. FIB step 1: platinum deposition. FIB step 2: platinum milling.

6. Mechanical Testing of the MEMS

Integrated MEMS require mechanical testing in addition to other forms of testing e.g. electrical, and can take the form of dimensional, dynamic and static tests. Optical techniques allow MEMS to be tested and modified precisely in a non-destructive manner. Figure 13 shows a schematic diagram of a workstation that contains a laser vibrometer for dynamic measurements, a surface profiler for static measurements and a Nd:Yag laser for laser ablation. Detailed operation specification of the workstation can be found in ref. 35.

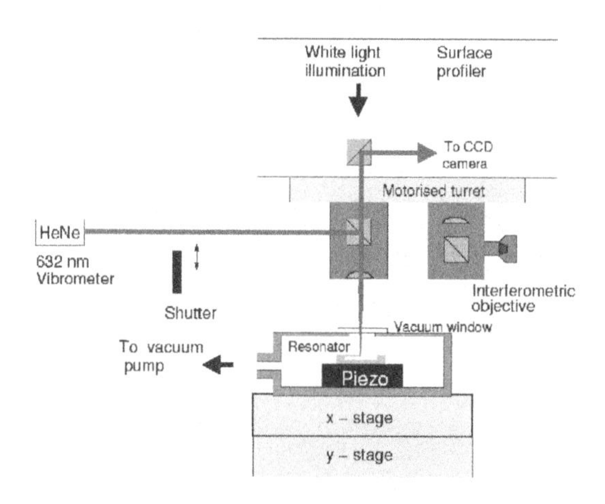

Figure 13. Optical workstation for resonator characterization.

7. Application Examples

Microelectromechanical systems based in silicon carbide, including accelerometer,[11,16] micro-motor,[36] pressure sensors,[21-25] gas sensors,[37,38] radiation detectors,[39] fuel atomizers,[40] have been demonstrated previously. For an excellent overview of the application areas of SiC, the reader is referred to chapter 5 of this book, which serves to illustrate in more detail the state-of-the-art SiC sensors, devices and systems that have been constructed so far.

Despite the large efforts devoted towards the research and development into SiC MEMS in the past decade, a lot remains to be done, especially in optimizing processes that are compatible with silicon, in order to reduce the cost of commercialization. A great impetus towards the future applications of SiC MEMS lies in its flexibility in integration with high temperature electronics. The possibility of MEMS and electronics that could operate at high temperatures and in harsh environments all in a single-chip module would no doubt enhance the commercialization prospects of MEMS in SiC.

8. Summary

This chapter has introduced the potential of using SiC as a MEMS material for harsh environments and has summarized the current research on SiC MEMS performed at Edinburgh. The future of SiC MEMS for harsh environments looks bright, especially when companies begin to commercialize SiC products[41] and as the understanding of the growth, etching and contact formation processes are advanced. The real potential of SiC MEMS applications will also be brought out when the integration of MEMS with electronics is proven.

REFERENCES

1. Hensler, R., Electronics.ca Publications, Research Report #GB270, BCC, (2002).
2. Mehregany, M., Zorman, C.A., Rajan, N. and Wu, C.H., *Proceedings of IEEE,* **86**, (1998), pp.1594-1610.
3. Sarro, P.M., *Sensors and Actuators,* **82**, (2000), pp.210-218.
4. Bouchaud, J. and Wicht, H., *Compound Semi & Microtechnology,* (2003), pp.26-29.

5. "Emerging Opportunities in Optical MEMS: 2003–2007", Communications Industry Researchers Inc., (2003).
6. Mirgorodsky, A.P., Smirnov, M.B., Abdelmounim, E., Merle, T. and Quintard, P.E., *Phys. Rev.*, **B52**, (1995), pp.3993-4000.
7. Fleishman, A.J., Zorman, C.A. and Mehregany, M., *J. Vac. Sci. Technol.*, **B16,** (1998), pp.536-539.
8. Pan, W.S. and Steckl, A.J., *J. Electrochem. Soc.*, **137,** (1990), pp.212-220.
9. Fleishman, A.J. *et al.*, *Proc. 9th Annual Int. Workshop on Microelectromechanical Systems*, San Diego, (1996), pp.473-478.
10. Fleishman, A.J., Wei, X., Zorman, C.A. and Mehregany, M., *Proc.7th Int. Conf. SiC III: Nitrides and Related Materials,* Sweden, (1997), pp.885-888.
11. Yasseen, A.A., Wu, C.H., Zorman, C.A. and Mehregany, M., *J. of Microelectromech. Systems*, **8**, (1999), pp.237-242.
12. Jiang, L., Cheung, R., Hassan, M., Harris, A.J., Burdess, J.S., Zorman, C.A. and Mehregany, M., *J. Vac. Sci. and Technol.,* **B21**, (2003), pp.2998-3001.
13. Jiang, L., Hassan, M., Cheung, R., Harris, A.J., Burdess, J.S., Zorman, C.A. and Mehregany, M., *Microelectronic Engineering*, **78-79**, (2005), pp.106-111.
14. Kuo, H.I., Zorman, C.A. and Mehregany, M., *Transducers '03: The 12th IEEE International, Conference on Solid-State Sensors, Actuators and Microsystems,* **2E86.P**, (2003), pp.742-745.
15. Shor, J.S., Goldstein, D. and Kurtz, A.D., *IEEE Transactions on Electron Devices,* **40**, (1993), pp.1093.
16. Wiser, R., Zorman, C.A. and Mehregany, M., *Transducers '03: The 12th IEEE International Conference on Solid-State Sensors, Actuators and Microsystems,* **3D3.5**, (2003), pp.1164-1167.
17. Roy, S., DeAnna, R.G., Zorman, C.A. and Mehregany, M., *IEEE Trans. on Electron Devices*, **49**, (2002), pp.2323-2332.
18. Yang, Y.T., Ekinci, K.L., Huang, X.M.H., Schiavone, L.M., Roukes, M.L., Zorman, C.A. and Mehregany, M., *Applied Physics Letters*, **78**, (2001), pp.162-164.
19. Okojie, R.S., Atwell, A.R., Kornegay, K.T., Roberson S.L., and Beliveau A., *Technical Digest of the 15th IEEE International Conference on MEMS*, Las Vegas, NV, (2002), pp.618-622.
20. Atwell, A.R., Okojie, R.S., Kornegay, K.T., Roberson S.L. and Beliveau A., *Nanotech 2002*, Puerto Rico, (2002).
21. Zappe S., Obermeier E., Möller H., Krötz G., Bonnotte E., Barriol Y., Decorps J.L., Rougeot C., Lefort O. and Menozzi G., *Conf. Transducers '99*, Sendai, Japan (1999), pp.346–349.
22. Zappe, S., Eickhoff, M. and Stoemenos, J., *Conf. Microelectronics, Microsystems and Nanotechnology,* Athens, Greece, (2000), pp.227-233.
23. Zappe, S., Franklin, J., Obermeier, E., Eickhoff, M., Moller, H., Krotz, G., Rougeot, C., Lefort, O. and Stoemenos, J., *Materials Science Forum*, **353-356** (2001), pp.753-756.
24. Eickhoff, M., Moller, H., Kroetz, G., Berg, J.V. and Ziermann, R., *Sensors and Actuators*, **A74**, (1999), pp.56-59.

25. Wu, C.-H., Stefanescu, S., Kuo, H.-I., Zorman, C.A. and Mehregany, M., *Conf. Transducers '01*, vol. 1, Munich, Germany (2001), pp.514-517.
26. Jiang, L., Cheung, R., Hedley, J., Hassan, M., Harris, A.J., Burdess, J.S., Zorman, C.A. and Mehregany, M., submitted to *J. Micromechanics and Microengineering*, (2005).
27. Huang, X.M.H., Zorman, C.A., Mehregany, M. and Roukes, M.L., *Transducers '03: The 12th IEEE International, Conference on Solid-State Sensors, Actuators and Microsystems,* **2E81.P**, (2003), pp.722-725.
28. Jiang, L., Cheung, R., Brown, R. and Mount, A., *Journal of Applied Physics,* **93**, (2003), pp.1376-1383.
29. Jiang, L., Plank, N.O.V. and Cheung R., *Microelectronic Engineering,* **67-68**, (2003), pp.369-375.
30. Nanse, G., Papirsr, E., Fioux, P., Moguet, F. and Tressaud, A., *Carbon*, **35**, (1997), pp.175-194.
31. Plank, N.O.V., Blauw, M.A., van der Drift, E. and Cheung, R., *Journal of Physics D; Applied Phys.*, **36**, (2003), pp.482-487.
32. Plank, N.O.V., Jiang, L., Gundlach A.M. and Cheung, R., *Journal of Electronic Materials,* **32**, (2003), pp.964-971.
33. Plank, N.O.V., Jiang, L., Gundlach, A.M. and Cheung, R., Proceedings of the 4th European Conference on Silicon Carbide and Related Materials, 2002, *Materials Science Forum*, **433-436**, (2003), pp.689-692.
34. Enderling, S., Jiang, L., Ross, A.W.S., Bond, S., Hedley, J., Harris, A.J., Burdess, J.S., Cheung, R., Zorman, C. A., Mehregany, M. and Walton, A. J., presented at *Nanotech 2004*, Boston, (2004).
35. Hedley, J., Harris, A., Burdess, J. and McNie, M., *Proceedings of the SPIE,* **4408** (2001), pp.402-408.
36. Yasseen, A.A., Wu, C.H., Zorman, C.A. and Mehregany, M., *IEEE Electron Device Letts*, **21**, (2000), pp.164-166.
37. Shields, V.B., Ryan, M.A., Williams, R.M., Spencer, M.G., Collins, D.M. and Zhang, D., *Inst. Phys. Conf. Ser.*, no 142, Ch. 7, (1996), pp.1067-1070.
38. Spetz, A.L., Tobias, P., Barabzahi, A., Martensson, O. and Lundstrom, I., *IEEE Trans. on Electron Devices,* **46**, (1999), pp.561-566.
39. Strokan, N.B., Ivanov, A.M., Savkina, N.S., Strelchuk, A.M., Lebedev, A.A., Syvajarvi, M. and Yakimova, R., *J. of Applied Physics*, **93**, (2003), pp.5714-5719
40. Rajan, N., Mehregany, M., Zorman, C.A., Stefanescu, S. and Kicher, T.P., *J. of Microelectromech. Systems*, **8,** (1999), pp.251-257.
41. www.FLXMicro.com

CHAPTER 2

DEPOSITION TECHNIQUES FOR SIC MEMS

Christian A. Zorman, Xiao-An Fu and Mehran Mehregany

Department of Electrical Engineering and Computer Science
Case Western Reserve University
Cleveland, Ohio 44106 USA
E-mail: caz@po.cwru.edu

SiC is a highly attractive material for microelectromechanical systems (MEMS) in applications where the material requirements call for a mechanically durable, chemically inert, and thermally stable semiconductor. Of the collection of materials that meet this requirement, SiC has risen to the forefront in large part because appropriate thin film deposition techniques are currently available and capable of producing MEMS-grade material in amorphous, polycrystalline and single crystal phases. This chapter reviews the major deposition techniques used to produce SiC for MEMS applications. These techniques include APCVD, LPCVD, PECVD, PVD and ion beam synthesis.

1. Introduction

From a historical perspective, microelectromechanical systems (MEMS) has largely been centered on the use of single crystal and polycrystalline silicon (polysilicon) as the principle structural materials, due to their favorable electrical and mechanical properties coupled with a mature infrastructure for production and processing borne from the IC industry. This combination has resulted in the emergence of MEMS as a viable enabling technology with significant potential for economic impact in areas that would benefit from miniaturized systems manufactured using batch processing techniques. A strong dependence on Si, however, restricts the application areas to those where the environmental conditions are relatively benign with respect to temperature, wear, radioactivity, corrosion and other "harsh" conditions. Many of these harsh environment applications, especially in the automotive and aerospace industries, are particularly strong candidates for MEMS technology, since the use of highly functional microsystems is considered a leading approach to significantly improve performance. To address this need and participate in these largely untapped application areas, researchers are developing deposition processes and micromachining techniques for materials that have the physical and chemical properties needed to withstand these harsh environments. Wide bandgap semiconductors such as SiC, diamond and GaN are particularly attractive because their large electronic bandgaps result in stable semiconducting properties at elevated temperatures. Of these, SiC is the leading material for MEMS due to its outstanding chemical and mechanical properties in conjunction with advanced deposition techniques that are available for single, polycrystalline and amorphous thin films. This chapter provides an overview of some of the leading methods being used to produce SiC films for MEMS applications. Because a review of research covering various aspects of SiC deposition techniques would alone constitute a rather large textbook, the examples cited in this chapter largely pertain to deposition techniques developed specifically for MEMS applications.

2. Deposition Issues Related to SiC for MEMS

From a processing perspective, polysilicon is a very convenient material for MEMS because electrically conductive, low-stress thin films are relatively straightforward to produce, thus setting a high standard for alternative materials like SiC. For most MEMS applications polysilicon is deposited in large, resistively headed furnaces that can accommodate large numbers (>100) of large area (150 mm dia.) substrates using silane as a precursor gas. The deposition temperatures are reasonably low, ranging from 580°C for amorphous films to 625°C for polycrystalline films. Polysilicon films nucleate readily and adhere very well to Si_3N_4 and SiO_2 substrate layers under a wide range of processing conditions.

While undoped polysilicon is highly resistive, high conductivity n- and p-type films can be produced using ion implantation, solid source diffusion, and in-situ doping techniques. The residual stress in as-deposited polysilicon films can either be tensile or compressive depending on the growth conditions, but can be greatly reduced by a post-deposition annealing step. These characteristics enable the fabrication of complex, multi-layer, microscale electromechanical devices using batch processing techniques.

By its very nature, SiC is a more difficult material to synthesize than polysilicon. Formation of SiC requires reactions between Si and C atoms under proper thermal and chemical conditions. Formation of stoichiometric SiC films generally requires temperatures above 700°C, with polycrystalline SiC (poly-SiC) requiring even higher substrate temperatures (> 800°C). Lattice and thermal mismatches between SiC and Si (the most common MEMS substrate) generally result in significant residual stresses in the SiC films. With respect to microstructure, SiC is thermodynamically stable to very high temperatures, thus limiting the use of annealing as a stress reduction technique to amorphous SiC films. The diffusion constants of nearly all elements in SiC are extremely low, rendering solid source diffusion impractical as a doping technique and leaving only ion implantation and *in situ* doping as options. For SiC films on Si wafers, implantation is also challenging in that the most effective processes require implant and/or annealing temperatures that approach the melting point of the substrate. In most cases, *in situ* doping is used to enhance the conductivity of SiC films, but because N is a shallow donor in SiC, issues related to atmospheric contamination and precursor gas purity make depositing semi-insulating and p-type material extremely challenging, especially in large-scale systems. Lastly, a processing furnace comparable in size and throughput to conventional polysilicon reactors has yet to become commercially available. In spite of these significant technical challenges, the attractiveness of SiC for MEMS has motivated substantial work in developing processes and techniques suitable for the production of MEMS-grade material.

In addition to the general differences in synthesis and processing, SiC and Si differ in another important aspect that differentiates the two materials: SiC is polymorphic. Currently the 3C-, 4H-, and 6H-SiC polytypes are the only polytypes that are relevant from a technological perspective due to the development of synthesis techniques for epitaxial layers and, in the case of 4H- and 6H-SiC, bulk substrates. 3C-SiC has emerged as the dominant polytype for MEMS applications due to the fact that it can be synthesized on Si wafers, whereas 6H-SiC and 4H-SiC form at temperatures above the melting point of Si. This is not to say that the hexagonal polytypes have no relevance to MEMS. Quite the contrary, bulk micromachined pressure[1] and acceleration sensors[2] have been successfully fabricated from 6H-SiC. In these examples, the devices were micromachined from commercially available 6H-SiC wafers on which epitaxial 6H-SiC films were grown using CVD techniques commonly employed in the fabrication of SiC microelectronics. As such, the key enabling technique was not the deposition technique, but rather the bulk etching process. Therefore, the deposition processes used to realize the 6H-SiC will not be reviewed here, and instead the chapter will focus on deposition techniques for 3C-SiC.

3. Atmospheric Pressure Chemical Vapor Deposition

3.1. Epitaxial 3C-SiC films

Among the first methods used to deposit SiC films for MEMS applications was atmospheric pressure chemical vapor deposition (APCVD). Throughout the 1980s and early 1990s, APCVD was the dominant method used to grow epitaxial SiC films (3C- and 6H-SiC) for electronic device applications. While most electronic-grade material is now grown by low–pressure chemical vapor deposition (LPCVD), APCVD is still being used for SiC MEMS. Because APCVD systems incorporate fewer temperature sensitive components than comparable low-pressure systems, very high substrate temperatures (> 1300°C) can easily be maintained at reasonable expense. This feature is particularly advantageous for SiC epitaxy, where temperatures typically range from roughly 1300°C for 3C-SiC on Si wafers to over 1700°C for 6H-SiC on

6H-SiC substrates. The key components of a typical APCVD reactor are shown in Figure 1. The reaction chamber consists of a double walled, fused silica tube with cooling water circulated between the walls. The cold wall design minimizes deposition on the walls of the chamber as well as particulate formation inside the reactor. Early systems, such as the one shown in Figure 1, were horizontally oriented with gas inlet and exhaust ports at opposing ends. To achieve the high deposition temperatures required for epitaxial growth, substrates are mounted onto a SiC-coated graphite susceptor that is positioned centrally within both the chamber and a current-carrying copper coil positioned just outside. The susceptor is heated by induction using a high power (kW) rf-generator connected to the copper coil. To minimize the effects of gas depletion across the substrate, the susceptor is tilted slightly with respect to the incoming gas flow.

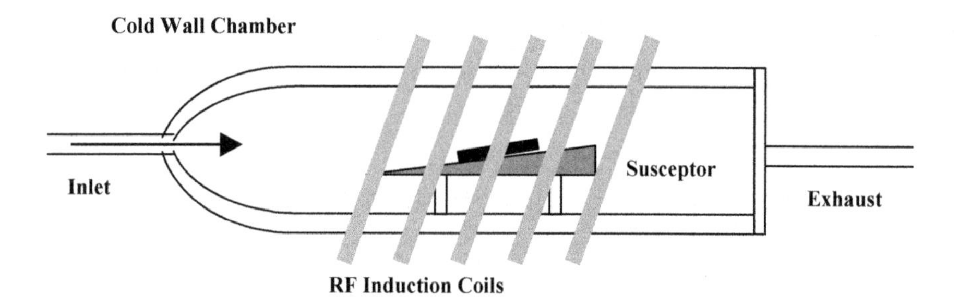

Figure 1. Schematic cross section of a horizontal APCVD reactor used for SiC epitaxy.

A reactor design quite similar to that shown in Figure 1 was used to grow some of the earliest SiC films evaluated for MEMS applications.[3] 3C-SiC films are particularly attractive for MEMS since the films can be readily grown on Si wafers and are impervious to all commonly used aqueous Si etchants. Such characteristics enable the use of Si bulk micromachining to fabricate 3C-SiC-based components. This is especially true for (100) Si, since anisotropic etchants such as KOH and TMAH can be used to form free standing membranes, cantilever beams and other such structures suspended over uniformly shaped cavities. Tong *et al.*[3] used such techniques to fabricate 3C-SiC thin film membranes for load-deflection testing and reported that the films had a Young's modulus over twice that of Si. Su *et al.*[4,5] used micromachined membranes and cantilevers to characterize the residual stress and Young's modulus of 3C-SiC and reported a Young's modulus as high as 794 GPa for Al-doped films.

Common processes for growing 3C-SiC on Si largely follow a process first detailed by Nishino *et al.*[6] and Powell *et al.*,[7] that is based on the conversion of a clean Si surface to 3C-SiC in a hydrocarbon gas, followed by 3C-SiC film growth using Si- and C-containing precursors. The process, shown schematically in Figure 2, is initiated with an in-situ etch of the Si surface at elevated temperature. This step serves to remove the native oxide from the Si surface and thus can be performed using either H_2 or HCl with the substrate heated to roughly 1000°C. After several minutes under these conditions, the substrate is usually cooled to below 500°C. A hydrocarbon gas mixed with a H_2 carrier gas is then introduced into the chamber and the susceptor is heated to roughly 1300°C. Various hydrocarbon gases have been used, the most common being C_3H_8. The temperature and flow conditions are maintained for a short period of time (90s to several minutes). During this period, the surface of the Si substrate is converted to 3C-SiC by reaction with the hydrocarbon gas in a process commonly called carbonization. Carbonization alone does not provide a practical means to grow 3C-SiC films for MEMS applications because the growth rates are nonlinear and decrease substantially with time. To produce films several microns in thickness, the growth rate is increased by adding a Si containing gas such as SiH_4 to the carrier gas. An adjustment of the hydrocarbon gas flow is often required in order to maintain a stoichiometric balance. Under the proper conditions, growth rates on the order of several microns per hour are readily achievable. 3C-SiC films can be doped with N or P to realize n-type conductivity and Al or B for p-type conductivity. Single crystalline 3C-SiC films can be heteroepitaxially grown using this method despite the nearly 20% lattice mismatch between 3C-SiC and Si. A representative TEM micrograph and corresponding electron diffraction pattern of an epitaxial 3C-SiC film is shown in Figure 3. The residual stresses in 3C-SiC films grown on Si substrates are nearly always tensile,[8] but work by Gourbeyre *et al.*[9] indicates that under certain carbonization conditions, compressive films can be realized.

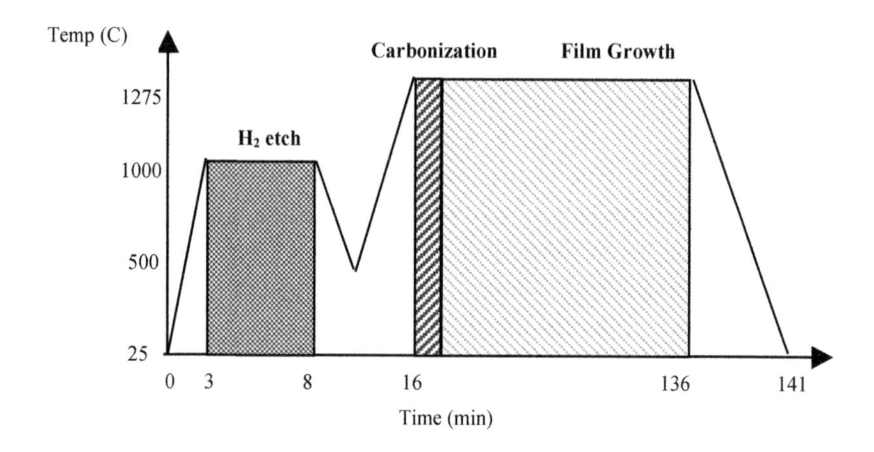

Figure 2. Schematic of a common carbonization-based 3C-SiC epitaxial growth process.

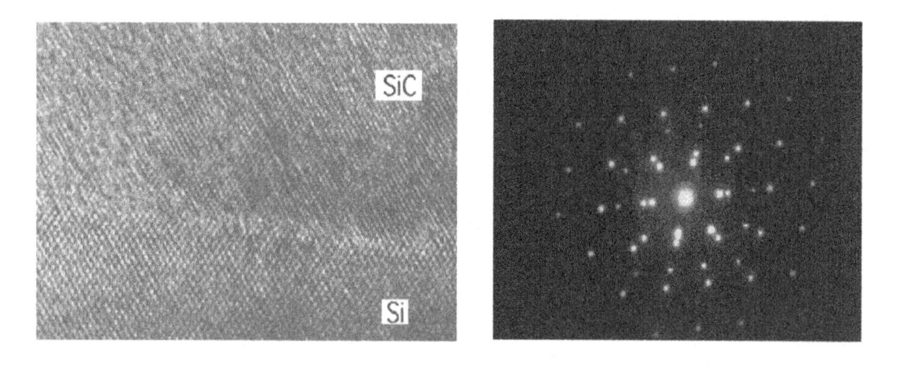

Figure 3. A high resolution transmission electron diffraction micrograph and a selected area diffraction pattern from a 3C-SiC film grown on a Si wafer by APCVD using a carbonization-based recipe.

Work by Tong *et al.*[3] and Mehregany *et al.*[8] showed that APCVD was a viable method to produce 3C-SiC films for Si bulk micromachining, but the reactors used in these studies were too small to demonstrate another potential attribute of MEMS fabrication, namely an adaptability to batch processing techniques by growth on large area substrates. To address this issue directly, Zorman *et al.*[10] developed a large scale APCVD furnace capable of holding 100 mm-diameter substrates. Figure 4 shows a schematic diagram of the system. In many respects, this reactor is identical to the common horizontal furnaces detailed previously with one significant difference; it incorporates a vertical geometry similar to reactors used for Si epitaxy. The vertical geometry minimizes particulate accumulation on the surface of the substrate, and more importantly, enables a second, 100 mm-diameter substrate to be loaded into the furnace. Epitaxial films from this reactor are of high quality and have been used in a wide range of single crystal 3C-SiC MEMS structures, including bulk micromachined membranes[11] and piezoresistive pressure sensors[12] as well as surface micromachined lateral resonators[13] and nanomechanical bridges.[14] Additionally, APCVD has been used to deposit thin 3C-SiC coatings on the active surfaces of bulk micromachined Si fuel atomizers to increase their resistance to erosion and to improve fuel atomization and spray angle.[15] Despite the poor conformality generally associated with APCVD processes, step heights of about 400 μm were completely coated.

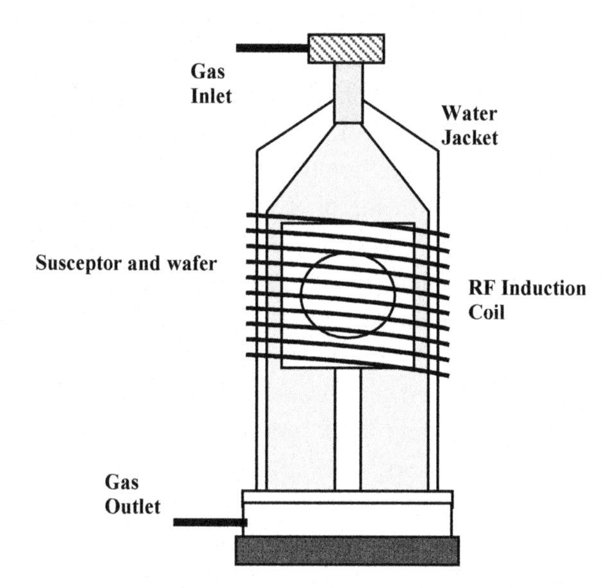

Figure 4. Schematic of a vertical APCVD reactor for growth of 3C-SiC on large area Si wafers.

3.2. Polycrystalline 3C-SiC films

Use of APCVD is not restricted to epitaxial processing, and in fact, APCVD recipes have been developed for polycrystalline 3C-SiC (poly-SiC) as well. In many respects, poly-SiC is more versatile than single crystalline 3C-SiC, since poly-SiC films can be deposited on a wide range of substrate materials, including SiO_2, polysilicon and Si_3N_4. The process used to deposit poly-SiC depends largely on the material characteristics of the substrate layer. For amorphous materials like SiO_2 and Si_3N_4, a single step process is used, since carbonization is neither effective nor desirable. The susceptor is simply ramped to the desired growth temperature and the precursors are injected into the reactor. The deposition temperature is lower than for 3C-SiC growth (1050°C to 1250°C) since epitaxy is not possible. Growth rates are comparable to epitaxial processes despite the temperature differences.

The ability to deposit poly-SiC films on SiO_2 and Si_3N_4 makes possible a surface micromachining process in SiC that parallels polysilicon. The Si_3N_4 layer is generally used to electrically isolate various components of an electromechanical device, while SiO_2 serves as a sacrificial layer on which the free standing structural components are fabricated. Various SiC MEMS components, such as lateral resonators[16] and micromotors[17] have been fabricated from poly-SiC films deposited on amorphous substrate layers.

SiC surface micromachining is not restricted to using SiO_2 sacrificial layers. From a materials perspective, the preferred substrate material for 3C-SiC film growth is Si. It follows, therefore, that polysilicon would be a viable sacrificial layer for poly-SiC MEMS. The first poly-SiC surface micromachined devices were, in fact, fabricated using polysilicon as a sacrificial layer.[18] A follow-on study showed that when the three step APCVD epitaxial growth process is performed on polysilicon, the resulting poly-SiC film takes on the microstructure of the polysilicon substrate.[19] In essence, when polysilicon is used as a sacrificial layer, the microstructure can be controlled by the growth conditions and substrate microstructure. Roy *et al.*[20] detailed the significance of this finding by showing that the mechanical properties of poly-SiC films grown on polysilicon relate to the microstructure of the films. When coupled with SiO_2 or Si_3N_4 for electrical isolation, poly-SiC and polysilicon form a powerful material system for surface micromachining, largely because the etchants used to selectively remove the polysilicon sacrificial layer do not attack poly-SiC, SiO_2 and Si_3N_4. As such, sacrificial release of free standing poly-SiC structures is straightforward. Numerous poly-SiC surface micromachined structures, such as kHz-frequency lateral resonators[21] and MHz-frequency vertical resonators[22] have been fabricated using polysilicon as a sacrificial layer.

4. Plasma Enhanced Chemical Vapor Deposition

Plasma enhanced chemical vapor deposition (PECVD) is an attractive deposition technique for SiC from a MEMS fabrication point-of-view because it enables SiC films to be deposited on a wide variety of substrate materials at temperatures much lower than APCVD, often between 200 and 400°C. In general, the microstructure of as-deposited PECVD SiC films is amorphous, but the films can be crystallized by annealing. Hydrogen is often incorporated in the films with a concentration that increases as the deposition temperature is decreased. The residual stress in PECVD SiC films is dependent on various deposition parameters, with stresses ranging from moderately compressive to moderately tensile.[23] Residual stress can be significantly reduced after film deposition by annealing.[24] Even before SiC became of interest as a structural material for MEMS, PECVD SiC was sometimes used as a masking material in Si bulk micromachining. This is because amorphous SiC films are generally resistant to KOH, HF and TMAH.[24]

Like most amorphous Si-derivative films, amorphous SiC can be deposited in commercial PECVD systems. Such a capability greatly enhances the widespread use of PECVD SiC in future MEMS applications since high throughput systems are readily available. A wide range of Si- and C-containing precursors can be used to deposit SiC by PECVD, with SiH_4 and CH_4[24,25] and $C_6H_{18}Si_2$ (hexamethyldisilane)[26] being some of the more common dual and single precursors specifically used for SiC MEMS. Like many PECVD processes, the recipes reported for PECVD SiC vary widely and depend on such parameters as the system manufacturer, concentration of precursor gases, substrate temperature, flow rates and the like.

The high degree of versatility in the PECVD process enables the incorporation of these films in a large variety of MEMS applications. Unlike 3C-SiC and poly-SiC which are used primarily as structural layers, PECVD SiC is more often used as a coating material for chemically sensitive components. Flannery *et al.*[25] illustrated several examples where PECVD SiC was used in Si-based microelectrode arrays, microfluidic channels, and piezoresistive pressure sensors.

Not surprisingly, PECVD SiC is a very versatile material to surface micromachine. As with poly-SiC, amorphous SiC can be deposited on sacrificial layers like SiO_2 and polysilicon. The much lower deposition temperatures also allow for materials such as Al and polyimide to be used as sacrificial layers.[27] Development of a polyimide-based surface micromachining process is particularly enabling, since the sacrificial etching step can be performed in an oxygen plasma.

5. Low Pressure Chemical Vapor Deposition

5.1. Overview

From a MEMS perspective, LPCVD is arguably the most versatile technique for production of SiC thin films. For starters, LPCVD is a practical method to produce 3C-SiC and poly-SiC films using a wide variety of precursors. The lower processing pressures result in a much higher degree of thickness uniformity on large-area substrates (up to 6 inch diameter) [28] than APCVD. Vacuum processing significantly reduces the incorporation of impurities from slowly outgassing reactor components. And while film growth rates tend to be much lower than for APCVD processes, LPCVD reactors generally accommodate more substrates, especially systems that utilize resistive heating. For these, and other technical reasons, a strong emphasis is currently being directed at the development of LPCVD processes for SiC MEMS.

5.2. Epitaxial 3C-SiC films

LPCVD-based processes for epitaxial 3C-SiC films emerged not long after the aforementioned three-step APCVD process became firmly established. As with the case with APCVD, the first LPCVD processes were designed to produce 3C-SiC films for electronic applications. In general, the processes were performed at high temperatures (> 1100°C) to insure growth of high quality material. Processes using dual[29,30] and single[31-33] Si- and C-containing precursors were developed. The dual precursor processes largely resembled APCVD in that carbonization steps were often incorporated, while the single precursors were used to push epitaxial growth to lower temperatures than could be achieved by APCVD and dual precursor LPCVD (e.g. > 1250°C), often without using carbonization. Both hot wall and cold wall reactors were used, and heating methods included induction heating,[30] resistive heating[29] and IR lamp heating.[31]

Development of LPCVD-based processes for SiC MEMS occurred at about the same that APCVD SiC films were being used in micromachined structures, most notably by Krotz *et al.*[32] This process utilized a mono-methylsilane (H_3Si-CH_3) and yielded 3C-SiC films on Si substrates at temperatures as low as 1000°C. This group used mono-methylsilane in multiple reactor configurations, including resistive heating[32] and IR lamp heating.[34] 3C-SiC films grown on Si wafers from mono-methylsilane proved to be particularly useful in device structures incorporating 3C-SiC free standing membranes, tethered plates and other Si bulk micromachined structures such as bridge-based flow sensors.[35] The authors were among the first to demonstrate the durability of SiC by showing that a 2 x 2 mm^2, 2 μm-thick 3C-SiC membrane pressurized to 1 bar and heated to 850°C did not undergo plastic deformation while an equivalent Si membrane was permanently deformed.[36] The films could be doped n-type using NH_3, and were particularly useful as high temperature piezoresistors. To address issues related to the difficulties in patterning 3C-SiC films on Si substrates, this group has developed growth processes using mono-methylsilane that are highly selective to Si surfaces relative to SiO_2,[37] and has used this process in conjunction with patterned SiO_2 films to fabricate 3C-SiC-based piezoresistive pressure sensors.[38]

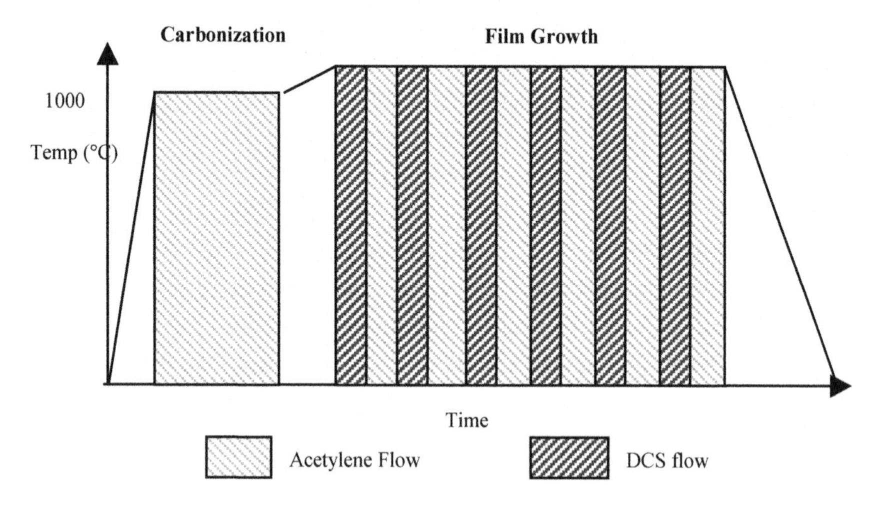

Figure 5. Schematic representation of the sequential deposition process used in atomic level epitaxy.

Dual precursors have also been used to grow epitaxial 3C-SiC films on Si substrates by LPCVD. In an example notable for its approach, Nagasawa *et al.*[29] studied the growth of 3C-SiC films using dichlorosilane (SiH_2Cl_2 or DCS) and acetylene (C_2H_2) over a temperature range of 750°C to 1050°C using a large scale, resistively heated LPCVD furnace. A multi-step growth process that included an acetylene-based carbonization step was used. Instead of simultaneously flowing DCS and acetylene during the film growth step, a sequential process was used, such that at any time only one precursor gas was flowing. In essence, the process involved the deposition of a thin epitaxial Si layer from DCS followed by the complete carbonization of this layer by acetylene. Film growth was continued in this manner until the desired film thickness was achieved. A schematic of the process, called atomic level epitaxy, is shown in Figure 5. Using this method, stoichiometric SiC films were deposited at temperatures above 800°C. The growth rate of the SiC films increased with increasing growth temperature for temperatures above 900°C, while a constant growth rate was observed between 800°C and 900°C. Single-crystal 3C-SiC was grown on (111) Si wafers at temperatures above 1000°C. In a follow-on study, the authors reported success in growing (100) 3C-SiC on (100) Si wafers.[39] Using the load-deflection technique on 1 μm-thick, bulk micromachined membranes, the authors found that the Young's modulus was around 450 GPa, with a mechanical burst strength of 0.2 kgf/cm^2.[39]

5.3. Polycrystalline 3C-SiC films

Polycrystalline 3C-SiC has become the dominant crystalline phase in SiC MEMS owing to its versatility in surface and bulk micromachining applications. As a result, LPCVD is rapidly emerging as the leading technique for the deposition of these films. Deposition of poly-SiC films by LPCVD basically requires C- and Si-containing precursors that dissociate and react at the desired substrate temperature. There exists an extensive body of research in the development of the appropriate precursors for poly-SiC growth by LPCVD, largely for applications other than MEMS. The list of precursors includes single precursors such as methyltrichlorosilane[40] (CH_3SiCl_3), tetramethylsilane[41] ($Si(CH_3)_4$), 1,3-disilabutane[42] (SiH_3-CH_2-SiH_2-CH_3), silacyclobutane[43] ($C_3H_6SiH_2$) and trimethylsilane[43] ($Si(CH_3)_3$), as well as dual precursors such as silane (SiH_4) with acetylene (C_2H_2),[44] disilane (Si_2H_6) with acetylene,[45] and dichlorosilane (SiH_2Cl_2) with acetylene.[46]

Tetramethylsilane (TMS) is a popular single source precursor due to its ease in handling and thus is of particular interest for use in commercial deposition systems. In a series of studies relevant to SiC MEMS, Clavaguera-Mora *et al.*[41,47] characterized poly-SiC films deposited by LPCVD using TMS at temperatures from 900°C to 1150°C on thermally oxidized Si wafers. As shown in the XRD micrographs in Figure 6, preferentially oriented, (111) 3C-SiC poly-SiC films were obtained at temperatures below 1000°C, while (200) 3C-SiC oriented films were deposited at 1150°C. TEM micrographs (not shown) indicated that at 1080°C the growth was columnar while at 1130°C, the growth was characterized by the formation of small equiaxed grains. The authors also reported that surface roughness and preferential orientation increased with increasing film thickness, deposition temperature and TMS-to-H_2 flow ratio. Figure 7 shows the influence of deposition temperature on the residual stress in the as-deposited films. The columnar films deposited below 1100°C exhibited compressive stresses, while the equiaxed films deposited at 1130°C had tensile stresses, with the film thickness and TMS flow rate influencing the magnitude of the residual stress. The contrasting microstructures and the incorporation of impurities were speculated to be the origin of the observed differences in stress.

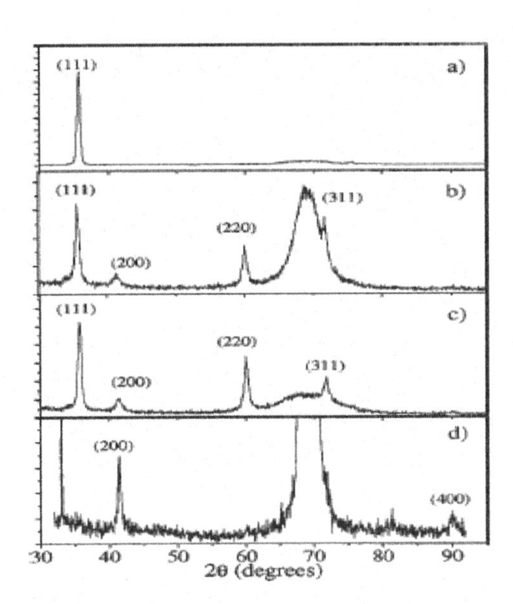

Figure 6. X-ray diffraction spectra for SiC films grown by LPCVD using TMS at: (a) 1080°C, (b) 1105°C, (c) 1130°C, and (d) 1150°C. The H_2-to-TMS ratio was 100.[47]

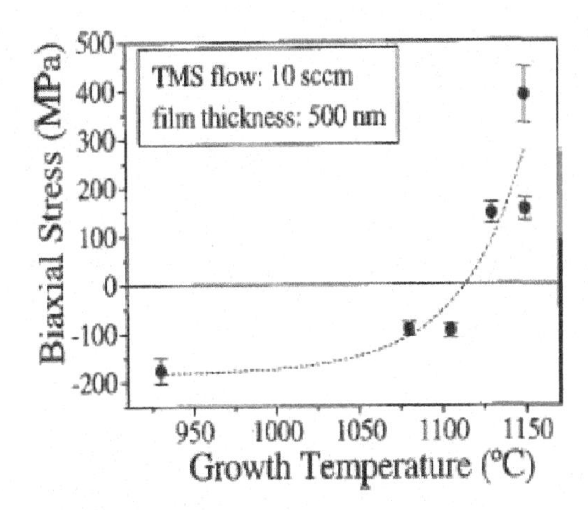

Figure 7. Residual stress versus deposition temperatures for poly-SiC films deposited by LPCVD using TMS as a precursor.[47]

While initial efforts at developing disilabutane (DSB) were for more traditional SiC applications, Stoldt *et al.*[48,49] were the first to extend the use of DSB directly to MEMS by demonstrating that poly-SiC films could be deposited using DSB on both planar Si substrates and prefabricated polysilicon MEMS devices. Pin-hole free, stoichiometric poly-SiC films were deposited at temperatures as low as 800°C using DSB. As expected, SiC films deposited at higher temperatures exhibited a larger average grain size than films deposited at the lower temperatures. The DSB-based process was successful in coating all the exposed surfaces of released polysilicon lateral resonant structures including the bottom surfaces, and the SiC-coated devices were much stiffer and more resistant to chemical attack than uncoated devices. Figure 9 shows SEM images of a lateral resonator following the deposition of a 35 nm-thick SiC coating at 800°C.

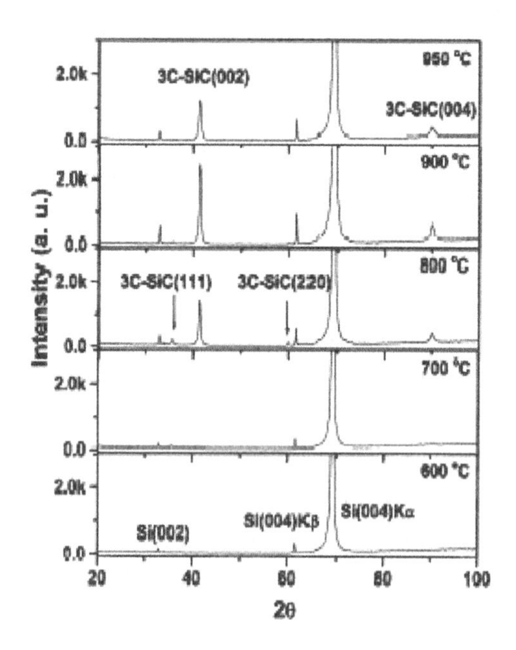

Figure 8. X-ray diffraction spectra from SiC thin films grown on Si (100) by LPCVD using 1,3 disilabutane at various deposition temperatures.[48]

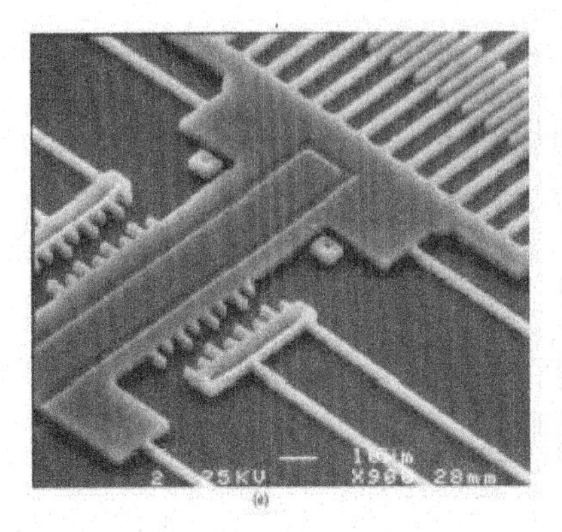

Figure 9. SEM images of a released poly-Si lateral resonator coated with a thin SiC film: (a) top view of resonator; (b) close-up of the resonator sidewalls.[49]

Acetylene is arguably the most popular C-containing gas used in dual precursor LPCVD poly-SiC processes. In an effort to identify appropriate conditions for producing bulk micromachined SiC X-ray mask membranes, Murooka *et al.*[50] investigated the stress, optical transparency, and surface roughness of SiC films deposited using SiH_4, C_2H_2, and HCl. The dependence of stress on the C-to-Si ratio in the source gas was affected by the addition of HCl, and a low stress, highly transparent membrane was obtainable by adjusting these parameters. The group found that the best films were deposited at 1050°C and a pressure of 1 kPa. In a second study, this group found that much like the addition of HCl in polysilicon LPCVD, the growth rate of poly-SiC was well described by a model that included the nucleation of particles in the gas phase, a change in the reaction species from hydrides to chlorides, and etching or inhibition of SiC growth by HCl.[44]

Recently, Zorman *et al.*[51] reported a unique LPCVD furnace designed specifically for high volume production of poly-SiC films for SiC MEMS applications. The furnace is comparable in size to conventional polysilicon LPCVD furnaces, measuring 2007 mm in length and 225 mm in diameter and capable of holding up to 100, 150 mm-diameter substrates. The system is resistively heated and utilizes SiH_2Cl_2 and C_2H_2 as precursors. Poly-SiC films have been deposited at temperatures ranging from 750°C to 900°C and pressures ranging from 460 mTorr to 5 Torr. Stoichiometric SiC films were deposited at temperatures between

800°C and 900°C at these pressures, and the non-stoichiometric films were carbon-rich and amorphous. XRD analysis showed that the stoichiometric poly-SiC films exhibit a strong (111) 3C-SiC texture. A schematic of the furnace is shown in Figure 10.

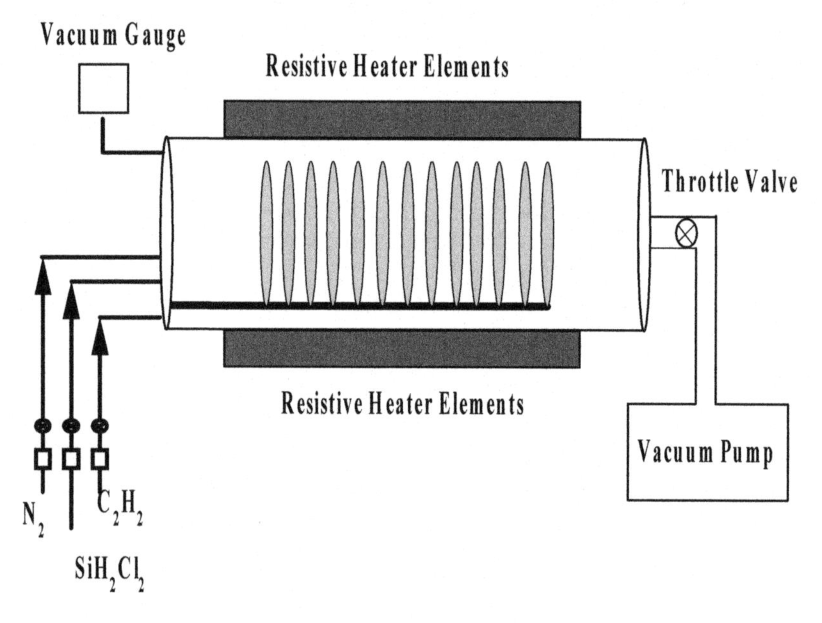

Figure 10. Schematic diagram of a high throughput LPCVD poly-SiC furnace.

The residual stresses in the films exhibited a strong dependence on the deposition pressure and varied from roughly 740 MPa at 460 mTorr to –75 MPa at 5 Torr, with very low stresses measured in films deposited around 2.5 Torr. Single-layer cantilever beams (Figure 11) fabricated from the low stress films indicated that the films also had a negligible stress gradient. From a mechanical perspective, the films exhibit the properties required of a material for use in surface micromachined devices.

Figure 11. Poly-SiC cantilevers fabricated from a low stress film deposited using DCS and acetylene by LPCVD at 900°C and 2.5 Torr.

Deposition of poly-SiC films by LPCVD at low tempertures is not limited to the use of acetylene as a C-containing precursor. Recently, Behrens *et al.*[52] reported the use of silane and carbontetrabromide (CBr_4) as precursors for LPCVD of poly-SiC. Stoichiometric films were deposited at a temperature of 940°C at growth rates of 250 nm/hr. The films provided excellent protection of Si substrates during bulk micromachining processes and were easily fabricated into free-standing membranes for load-deflection testing.

6. Doping of LPCVD Poly-SiC Films

For many MEMS applications, the electrical properties of the structural films are critically important. Unlike polysilicon (which is relatively easy to dope), the chemical inertness of SiC limits doping methods to ion implantation and *in situ* doping. From a processing perspective, *in situ* doping is the preferred method as it involves fewer processing steps, does not lead to excessive damage of the material and is relatively easy to incorporate into LPCVD processes. N-type conductivity via nitrogen doping is easy to achieve in 3C-SiC films since the processing temperatures are high and the crystal quality is sufficient to realize reasonably high electrical conductivities. In fact, the conductivity in 3C-SiC films grown by APCVD is generally suitable for electrostatic MEMS devices using only residual nitrogen in the deposition system.[13] For reproducible results, dopants such as N_2 and NH_3 are added to the gas flow during the film growth step.

Doping of poly-SiC films is more challenging than single crystal films primarily because doping in SiC is strongly dependent on processing temperature. Low deposition temperatures lead to entrapment of dopants in grain boundaries as a result of reduced adatom surface mobilities. Entrapment significantly reduces the efficiency of the doping process, resulting in films with relatively high resistivities as compared with single crystal films grown using the same dopant gas flows. Since the diffusion constants of impurities in SiC are extremely low, redistribution of dopants in poly-SiC by annealing is not a practical option, especially when Si is used as a substrate. In addition, the grain size of poly-SiC is generally dependent on substrate temperature, with smaller grained films deposited at lower temperatures. Grain growth by a post deposition annealing step is not practical for films deposited on Si, and might also lead to unwanted changes in residual stress. These, and other effects, make doping of poly-SiC a challenging task.

Several efforts are underway to develop robust *in situ* doping processes for poly-SiC films. Wijesundara *et al.*[53] recently reported low temperature doping of poly-SiC films by LPCVD using NH_3 as a dopant and 1,3 disilabutane as the precursor. Figure 12 shows the resistivity and conductivity of the films as a function of NH_3 in the feed gas. The conductivity measurements indicated that successful doping of the SiC was achieved with 2% NH_3 content, yielding an electrical resistivity of approximately 0.02 Ωcm. The investigators found that increasing the NH_3 flow affected the growth rate, chemical composition, and the crystalline quality of the as-deposited film. Figure 12 also shows the SiC growth rate as a function of NH_3 concentration in the feed gas. A set of optimized conditions with respect to conductivity was identified such that the growth rate and crystal quality of the films were not significantly affected.

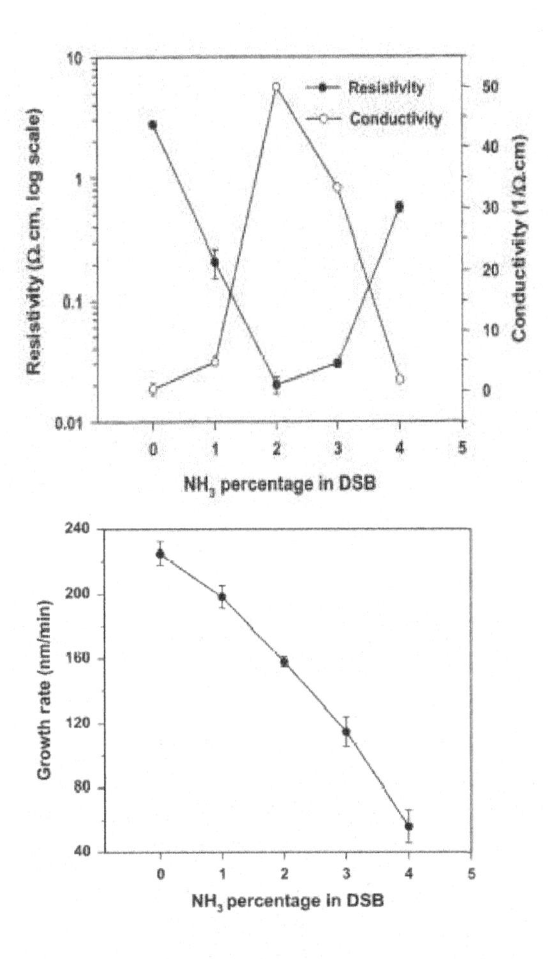

Figure 12. Resistivity and conductivity of poly-SiC films as a function of percentage of NH_3 in the feed gas (top), and SiC growth rate as a function of NH_3 percentage in the feed gas (bottom).[53]

In situ doping of poly-SiC films deposited from dual precursors has been successfully performed as well. Homma *et al.*[54] reported the development of SiC strain gauge-based sensors using p-type poly-SiC films deposited from SiH_4 and CH_4 precursors and doped with B using diborane (B_2H_6). The films were deposited by PECVD with the substrate resistively heated to 800°C to ensure that the films had a polycrystalline microstructure. The as-deposited films were predominantly (111) 3C-SiC and had a maximum piezoresistive gauge factor of 6. Doped poly-SiC films deposited on SiO_2 coated Si substrates were successfully used as piezoresistors in microfabricated pressure sensors.

Use of dopants in SiC is not limited to increasing the conductivity of the material. In an unanticipated finding, Murooka *et al.*[55] reported that the Young's modulus of poly-SiC can be increased considerably by doping the films with B_2H_6. The films were deposited in a resistively heated LPCVD reactor using SiH_4 and C_2H_2 precursors. In this case, bulk micromachined membranes were fabricated and tested using the load-deflection technique. The group found that the Young's modulus of undoped films was 480 GPa while that for boron doped films was 600 GPa for B_2H_6 flow rates such that B/Si = 0.02.

7. Other Deposition Methods

Although it is highly likely that CVD techniques will continue to play a dominant role in the production of thin films for SiC MEMS, alternative techniques will likely be employed in situations where the use of CVD is prohibited. One such niche application with considerable potential is the use SiC as a coating material on temperature sensitive substrates. For such applications, PVD methods are currently being explored. A principal advantage of PVD is that non-hydrogenated, amorphous SiC films can be deposited at room temperature on MEMS-type substrates, as recently shown by Ledermann *et al.*[56] This group has developed a magnitron sputtering process that can be used to deposit stress-controlled SiC films on planar and non-planar Si surfaces to thicknesses of several microns. The films were deposited in a commercial sputtering system using a stoichiometric SiC target and an Ar sputtering gas. Pinhole-free, low stress films resistant to standard KOH-based Si etchants were produced using this method.

Use of PVD SiC films is not limited to protective coating applications. In fact, sputtered SiC films have long been known to exhibit desirable electrical resistance characteristics at high temperatures, and were among the first of any type of SiC to be used in sensor applications. Nagai *et al.*[57] developed an rf sputtering technique to deposit poly-SiC films for high temperature thermistors. The films were deposited on alumina substrates at elevated temperature (650°C) from a sintered SiC target. Vacuum annealing at 900°C made the films suitable for thermistors designed to operate from 0 to 500°C.

Synthesis of thin SiC layers for MEMS applications is not restricted to conventional thin film deposition processes, as recently demonstrated by a technique reported by Serre *et al.*[58,59] This group has developed novel ion beam synthesis techniques to form 3C-SiC and poly-SiC on Si wafers in an effort to address issues related to residual stress and stress gradients. The processes involve implantation of C ions at moderate doses to form the SiC layer, sometimes performed at 500°C and sometimes followed by a high temperature anneal. To form 3C-SiC layers, implantation is performed into single crystal Si wafers, while poly-SiC layers are formed by implanting into LPCVD polysilicon films. Membranes, cantilever beams and other thin, single layer structures have been fabricated from these films.

8. Conclusions

A wide variety of deposition techniques have been developed to support the production of SiC films that are suitable for MEMS applications. With a few notable exceptions, SiC MEMS are fabricated from the 3C-SiC polytype, the only polytype that can be grown on Si wafers and at temperatures approaching those used in polysilicon processing. Processes now exist for single crystalline, polycrystalline and amorphous SiC thin films suitable for micromachining applications, as evidenced by the recent development of SiC-based pressure sensors, flow sensors, micromotors and other micromachined structures and devices. Progress has steadily been made to adapt these SiC deposition processes to commercial, large scale, high throughput systems. Film quality and process reproducibility in a few of these systems have evolved to the point that prototypes comparable to commercial polysilicon furnaces are beginning to emerge. For applications that are intolerant to conventional deposition approaches, specialized SiC deposition techniques have been developed. As SiC-related deposition technology continues to mature, SiC will undoubtedly find its way into many more MEMS applications.

REFERENCES

1. Okojie, R.S., Ned, A.A. and Kurtz, A.D., *Sen. Actuators,* **A66,** (1998), pp.200-204.
2. Okojie, R.S., Atwell, A.R., Kornegay, K.T., Roberson, S.L. and Beliveau, A., *Proc. 15th Int. Conf. Microelectromech. Sys.,* (IEEE, Piscataway NJ, 2002), pp.618-622.
3. Tong, L. and Mehregany, M., *Appl. Phys. Lett.,* **60,** (1992), pp.2992-2994.
4. Su, C.M., Fekade, A., Spencer, M. and Wuttig, M., *J. Appl. Phys.,* **77,** (1995), pp.1280-1283.
5. Su, C.M., Wuttig, M., Fekade, A. and Spencer, M., *J. Appl. Phys.,* **77,** (1995), pp.5611-5615.
6. Nishino, S., Powell, J.A. and Will, H.A., *Appl. Phys. Lett.*, **42,** (1983), pp.460-462.
7. Powell, J.A., Matus, L.G. and Kuczmarski, M.A., *J. Electrochem. Soc.,* **134,** (1987), pp.1558-1565.
8. Mehregany, M., Tong, L., Matus, L.G. and Larkin, D.J., *IEEE Trans. Elec. Dev.,* **44,** (1997), pp.74-79.
9. Gourbeyre, C., Chassange, T., LeBerre, M., Ferro, G., Gautier, E., Monteil, Y. and Barbier, D., *Sens. Actuators,* **A99,** (2002), pp.31-34.
10. Zorman, C.A., Fleischman, A.J., Dewa, A.S., Mehregany, M., Jacob, C., Nishino, S. and Pirouz, P., *J. Appl. Phys.,* **78,** (1995), pp.5136-5138.
11. Mitchell, J. S., Zorman, C.A., Kicher, T., Roy, S. and Mehregany, M., *J. Aerospace Eng.,* **16,** (2003), pp.46-54.
12. Wu, C.H., Stefanescu, S., Kuo, H.-I., Zorman, C.A. and Mehregany, M., *Tech. Dig. 11th Inter. Conf. Solid State Sen. Actuators* (IEEE, Piscataway, NJ, 2001), pp.514-518.
13. Stefanescu, S., Yasseen, A.A., Zorman, C.A. and Mehregany, M., *Proc. 10th Int. Conf. Solid State Sens. Actuators* (IEEE, Piscataway, NJ, 1999), pp.194-198.
14. Yang, Y.T., Ekinci, K.L., Huang, X.M.H., Schiavone, L.M., Roukes, M.L., Zorman, C.A. and Mehregany, M., *Appl. Phys Lett.,* **78,** (2001), pp.162-165.

15. Rajan, N., Zorman, C.A. and Mehregany, M., *Thin Solid Films,* **315,** (1998), pp.170-178.
16. Fleischman, A.J., Wei, X., Zorman, C.A. and Mehregany, M., *Mat. Sci. Forum,* **264-268,** (1998), pp.885-889.
17. Yasseen, A.A., Wu, C.H., Zorman, C.A. and Mehregany, M., *IEEE Elec. Dev. Lett.,* **21,** (2000), pp.164-166.
18. Fleischman, A.J., Roy, S., Zorman, C.A. and Mehregany, M., *Proc. 9th Int. Wrkshp. Microelectromech. Sys.* (IEEE, Piscataway, NJ, 1996), pp.234-238.
19. Zorman, C.A., Roy, S., Wu, C.H., Fleischman, A.J. and Mehregany, M., *J. Mat. Res.,* **13,** (1998), pp.406-412.
20. Roy, S., Zorman, C.A. and Mehregany, M., *Mat. Res. Soc. Symp. Proc.,* **657,** (2000), EE9.5.1 – EE9.
21. Roy, S., DeAnna, R.G., Zorman, C.A. and Mehregany, M., *IEEE Trans. Elec. Dev.,* **49,** (2002), pp.2323-2332.
22. Wiser, R.F, Zorman, C.A. and Mehregany, M., *Tech. Dig. 12th Int. Conf Solid State Sen. Actuators and Microsys.* (IEEE, Piscataway, NJ, 2003), pp.1164-1167.
23. Sarro, P., *Sens. Actuators,* **82,** (2000), pp.210-218.
24. Sarro, P., deBoer, C.R., Korkmaz, E. and Laros, J.M.W., *Sen. Actuators,* **A67,** (1998), pp.175-180.
25. Flannery, A.F., Mourlas, N.J., Storment, C.W., Tsai, S., Tan, S.H., Heck, J., Monk, D., Kim, T., Gogoi, B. and Kovacs, G.T.A., *Sens. Actuators,* **A70,** (1998), pp.48-55.
26. Klumpp, A., Schaber, U., Offereins, H.L., Kuhl, and Sandmaier, H., *Sens. Actuators,* **A41-42,** (1994), pp.310-316.
27. Bagolini, A., Pakula, L., Scholtes, T.L., Pham, H.T.M., French, P.J. and Sarro, P.M., *J. Micromech. Microeng.,* **12,** (2002), pp.385-389.
28. Yagi, K. and Nagasawa, H., *Mat. Sci. Forum,* **264-268,** (1998), pp.191-194.
29. Nagasawa, H. and Yamaguchi, Y., *Thin Solid Films,* **225,** (1993), pp.230-234.
30. Eshita, T., Suzuki, T., Hara, T., Mieno, F., Furumura, Y., Maeda, M., Sugii, T. and Ito, T., *Mat. Sci. Res. Symp. Proc.,* **116,** (1988), pp.357-362.
31. Madapura, S., Steckl, A.J. and Loboda, M., *J. Electrochem. Soc.,* **146,** (1999), pp.1197-1202.
32. Krotz, G., Legner, W., Muller, G., Grueninger, H.W., Smith, L., Leese, B., Jones, A. and Rushworth, S., *Mat. Sci. Eng.,* **B29,** (1995), pp.154-159.
33. Golecki, I., Reidinger, F. and Marti, J., *Appl. Phys. Lett.,* **60,** (1992), pp.1703-1705.
34. Moller, H., Legner, W. and Krotz, G., *Mat. Sci. Forum,* **264-268,** (1998), pp.171-174.
35. Lyons, C., Freidberger,A., Wesler, W., Muller, G., Krotz, G. and Kassing, R., *Proc. 11th Int. Wrkshp. Microelectromech. Sys.* (IEEE, Piscataway, NJ, 1997), pp.356-360.

36. Krotz, G., Legner, W., Moller, H., Sonntag, H. and Muller, G., *Proc. 8th Int. Conf. Solid State Sen. Actuators – Eurosensors IX* (IEEE, Piscataway, NJ, 1995), pp.186-189.
37. Eickhoff, M., Moller, H., Rapp, M. and Krotz, G., *Thin Solid Films,* **345,** (1999), pp.197-199.
38. Eickhoff, M., Moller, H., Krotz, G., v. Berg, J. and Ziermann R., *Sen. Actuators,* **74,** (1999), pp.56-59.
39. Yamaguchi, Y., Nagasawa, H., Shoki, T. and Annaka, N., *Proc. 8th Int. Conf. Solid State Sen. Actuators – Eurosensors IX* (IEEE, Piscataway, NJ, 1995), pp.190-193.
40. Lee, K.S., Park, J.Y., Kim, W.J. and Hong, G.W., *J. Mater. Sci. Lett.,* **20,** (2001), pp.1229-1231.
41. Clavaguera-Mora, M.T., Rodriguez-Viejo, J., El Felk, Z., Hurtos, E., Berberich, S., Stoemenos, J. and Clavaguera, N., *Dia. Rel. Mat.,* **6,** (1997), pp.1306-1310.
42. Boo, J.H., Lee, S.B., Yu, K.S., Sung, M.M. and Kim, Y., *Surf. Coat. Tech.,* **131,** (2000), pp.147-152.
43. Chen, J., Scofield, J. and Steckl, A.J., *J. Electrochem. Soc.,* **147,** (2000), pp.3845-3849.
44. Murooka, K., Higahikawa, I. and Gomei, Y., *J. Crys. Growth,* **169,** (1996), pp.485-490.
45. Hong, L., Shimogaki, Y., Egashira, Y. and Komiyama, H., *J. Electrochem. Soc.,* **139, (**1992), pp.3652-3659.
46. Wang, C.F. and Tsai, D.S., *Mat. Chem. Phys.,* **63,** (2000), pp.196-201.
47. Hurtos, E. and Rodriguez-Viejo, J., *J. Appl. Phys*., **87,** (2000), pp.1748-1758.
48. Stoldt, C.R., Carraro, C., Ashurst, W.R., Gao, D., Howe, R.T. and Maboudian, R., *Sens. Actuators,* **A97,** (2002), pp.410-415.
49. Stoldt, C.R., Fritz, M.C., Carraro, C. and Maboudian R., *Appl. Phys. Lett.,* **79,** (2001), pp.347-349.
50. Murooka, K., Itoh, M., Komano, H. and Gomei, Y., *Jap. J. Appl. Phys.,* **30,** (1991), pp.3074-3077.
51. Zorman, C.A., Rajgopal, S., Fu, X.A., Jezeski, R., Melzak, J. and Mehregany, M., *Electrochem. Solid State Lett.,* **5,** (2002), G99-G101.
52. Behrens, I., Peiner, E., Bakin, A.S. and Schlachetzki, A., *J. Micromech. Microeng.,* **12,** (2002), pp.380-384.
53. Wijesundara, M.B.J., Stoldt, C.R., Carraro, C., Howe, R.T. and Maboudian, R., *Thin Solid Films,* **419,** (2002), pp.69-75.
54. Homma, T., Kamimura, K., Cai, H.Y., Onuma, Y., *Sens. Actuators,* **A40,** (1994), pp.93-96.
55. Murooka, K., Higashikawa, I. and Gomei, Y., *Appl. Phys. Lett.,* **69,** (1996), pp.37-39.
56. Ledermann, N., Baborowski, J., Muralt, P., Xantopoulos, N. and Tellenbach, J.M., *Surf. Coat. Tech.,* **125,** (2000), pp.246-250.
57. Nagai, T. and Itoh, M., *IEEE Trans. Ind. Appl.,* **26,** (1990), pp.1139-1143.

58. Serre, C., Perez-Rodriguez, A., Romano-Rodriguez, A., Morante, J.R., Esteve, J. and Acero, M.C., *J. Micromech. Microeng.*, **9,** (1999), pp.190-193.
59. Serre, C., Perez-Rodriguez, A., Morante, J.R., Esteve, J., Acero, M.C., Kogler, R. and Skorupa, W., *J. Micromech. Microeng.*, **10**, (2000), pp.152-156.

CHAPTER 3

REVIEW OF ISSUES PERTAINING TO THE DEVELOPMENT OF CONTACTS TO SILICON CARBIDE: 1996–2002

Lisa M. Porter and Feroz A. Mohammad

Department of Materials Science and Engineering
Carnegie Mellon University, Pittsburgh, PA 15213
E-mail: lporter@andrew.cmu.edu

1. Introduction

Since the time of writing a comprehensive review of SiC contacts in 1995,[1] the work in this field has flourished. Papers pertaining to contacts to SiC that were published in the seven years 1996–2002 number at least 250% more than those published in all years prior to 1996. Whereas this trend reflects the successful growth in the research and development of SiC devices, the volume of literature makes it unfeasible to attempt a comprehensive review of all papers pertaining to SiC contacts. Instead, we have chosen to focus on the following four topics that we deem are important and timely: (1) the thermal stability of ohmic and rectifying contacts; (2) ohmic contacts to p-type SiC; (3) ohmic contacts using nickel and (4) the effects of defects on contacts. This chapter is organized into sections that are devoted to each of the four topics.

2. Thermal Stability

2.1. Thermal Stability of Ohmic Contacts

The intrinsic properties of SiC give it substantial advantages over Si for use in high temperature electronics. Therefore, there is strong interest in developing SiC-based devices for both military and civilian high-temperature electronics such as high-energy, pulsed-power electronics for electric weapon technology[2] and gas sensors that can be used at high temperatures (~300–800°C) and in aggressive environments (e.g. for the control of industrial and automobile emissions).[3]

However, while the intrinsic, bulk properties of this wide band gap semiconductor allow for the expansion of the current limits on operating temperature, interfaces between dissimilar materials at the discrete device level as well as at the packaging and interconnection levels often interact at relatively low temperatures. Thus, it is important to identify and select phases that are thermodynamically stable with each other at these temperatures.

Park *et al.*[4] classified metal (M)-SiC reactions according to two types based on characteristics of the isothermal M-Si-C phase diagrams. The first type includes metals that do not form binary carbides, and the second type includes metals that form at least one binary carbide. The two types of systems were generalized by the following chemical reactions:

Type 1: M + SiC —> silicides + C
Type 2: M + SiC —> silicides + carbides
or M + SiC —> silicides + carbides + $M_xSi_yC_z$

Interestingly, although SiC is stable at high temperatures, has strong interatomic bonding and is chemically inert, its free energy of formation is relatively high. Therefore, it is unstable to reaction with most metals; i.e. it is thermodynamically favorable for SiC to react with metals to form silicides and/or carbides at elevated temperatures.

Accordingly, some groups have performed thermodynamic calculations to determine which phases should form when certain metals react with SiC.[5,6] For example, W-Si-C ternary phase diagrams were calculated using the Gibbs free energy from tabulated thermodynamic data[6]. For temperatures below ~700°C, WSi_2 and WC are stable with SiC. For temperatures between ~700 and 1870°C, W_5Si_3 and WC are the predicted, stable phases when W reacts with SiC. Above ~1870°C, W_2C is stable instead of WC. While WSi_2 should not form from the deposition of W on SiC at temperatures above ~700°C, WSi_2 is itself stable with SiC at temperatures betweeen R.T. and 1900°C when carbide phases are not present.

According to Delucca and Mohney[7] Re is the only transition metal that is stable with SiC. In this study they investigate M-Si-C phase equilibria and discuss trends that can be applied to thermally stable contacts. Thus, the study is useful for selecting phases that may be used for thermally stable contacts to SiC. For group IVB and VB metals all of the metal disilicides and mono-carbides are in equilibrium with SiC. In some cases other silicides and/or ternary phases are also in equilibrium with SiC. In contrast, the late transition metals (e.g. Co and Ni) form only silicides. Four Ni silicides (Ni_2Si, Ni_3Si_2, NiSi and $NiSi_2$) and two Co silicides (CoSi and $CoSi_2$) are in equilibrium with SiC.

An example of the application of these concepts is the work of Jang *et al.*[8] In this study TaC was selected because of its relatively low work function, its high melting point and its predicted, thermodynamic stability with SiC. This stability is represented by the tie line between TaC and SiC in the ternary phase diagram[9] shown in Figure 2.1. As predicted, cross-sectional TEM analyses showed no reaction between the TaC and SiC after annealing at 1000°C for 15 minutes to form the ohmic contacts (Figure 2.2).[8]

Similarly, WSi_2 and $MoSi_2$ were investigated[10] because they have stable tie lines with SiC up to at least 1800°C (for WSi_2) and 1600°C (for $MoSi_2$). Contacts that were off stoichiometry ($WSi_{1.8}$ and $MoSi_{2.5}$) displayed much poorer stability to annealing at 400°C than the stoichiometric films. These results indicate that even relatively small fluctuations from the stoichiometrically stable phases can make significant differences in the resulting properties. Other phases that are predicted to be stable with SiC and that have been investigated as thermally-stable ohmic contacts include TiC,[11,12] Ti_3SiC_2[13] and PtSi.[14]

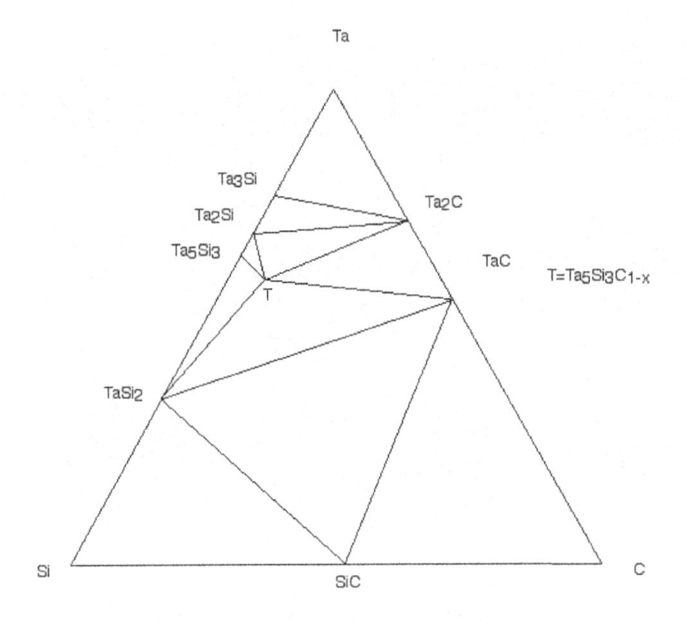

Figure 2.1. The approximate phase diagram for Ta-Si-C at 1000 °C. For a more accurate diagram see reference 9.

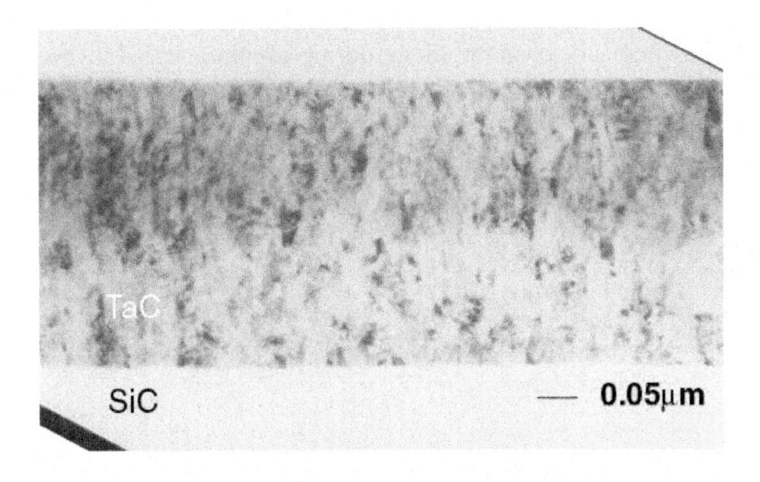

Figure 2.2. Cross-sectional TEM image of the TaC/6H-SiC interface after annealing at 1000°C for 15 min. (first published in reference 8). The image shows an abrupt interface with no sign of a reaction.

While thermodynamic calculations show that there are many phases that are in equilibrium with SiC at processing or potential operating temperatures, the additional, practical requirements for high-temperature contacts severely complicate the task of developing high-temperature metallization schemes. For example, unless the device is hermetically sealed, the contacts must be highly resistant to oxidation. Other requirements include sufficient electrical conductivity, the ability to be connected to the outside world (e.g. wire bonding), and appropriate interfacial electrical properties (e.g. contact resistance or Schottky barrier height). In order to meet all of these requirements, multiple layers are typically combined. The additional layers introduce thermodynamic instability to what may have originally been a stable metal/SiC interface (e.g. TaC/SiC). Therefore, one must consider diffusion barriers, which substantially slow the kinetics of diffusion/reaction between adjacent layers.

Table 1 summarizes various ohmic contact metallization schemes that have been investigated as a function of "long-term" annealing and/or elevated measurement temperature. A variety of materials, such as Cr and TiN, have been investigated as diffusion barriers. However, all have displayed some evidence of changes in the electrical properties or interfacial chemistry or morphology during the time scales investigated, unless high vacuum conditions were maintained to reduce oxidation of the contacts.

Table 1. Summary of results pertaining to the thermal stability of ohmic contacts to SiC. The term "long-term anneal" is used to differentiate from the annealing conditions used to form the ohmic contact. *Note: Multi-layered contacts are designated with slashes to separate the distinct layers; layers at the surface to the interface with SiC proceed from left to right. The metal on the top is listed first, and the SiC substrate is inferred last, e.g. Au/NiCr/SiC.

Metallization*	Substrate or Epilayer Polytype	Doping Type	Doping Level (cm^{-3})	'Long-term' Annealing Conditions	ρ_c (Ω cm^2) before 'long-term' anneal	ρ_c (Ω cm^2) after 'long-term' anneal	ρ_c (Ω cm^2) at elevated temperature	Phases Formed	Ref
Si/Co	6H	p	2×10^{19}	1100 °C for 3 h in vacuum	3.6×10^{-6}	Increased	1.5×10^{-5} @ 200 °C	$CoSi_2$	[*19*]
1) ITO (90% In_2O_3 – 10% SnO_2) 2) Ni-Cr 3) In	β-SiC/Si	n	5×10^{16} –3.7 x 10^{17}	No anneal; deposited at 200 °C	1) 5.6×10^{-1} 2) 7.8×10^{-2} 3) 6.6×10^{-2}	——	1) Unmeasurable at 300 °C 2) 7.2×10^{-3} at 300 °C 3) Unmeasurable at 300 °C	N.R.	[*20*]
W/Cr/Ni W/Mo/Cr	6H and 4H	n	10^{17}–10^{18}	650 °C for 1000 h in vacuum	5×10^{-6} 10^{-4}–10^{-2}	——	——	Ni_2Si and TiC Cr_3C_2	[*21*]
Al/Mo/Pt/Ti/Cr	3C	n	3×10^{18}	550 °C for 400 h in air	6×10^{-7}	7×10^{-5}	——	N.R.	[*22*]
Ni:Cr (80:20 wt.%)	6H and 4H	p	3.2×10^{17} and 1.4×10^{18} (6H); 4.8×10^{17} and 1.3×10^{19} (4H)	300 °C for 2500 h in vacuum	2.5×10^{-6} – 9.1×10^{-5} (6H); 1.2×10^{-5} – 1.5×10^{-4} (4H)	——	——	N.R.	[*23*]

CrB_2	6H	p	1.3×10^{19}	(1) 1100 °C/3.5 h + 1200 °C/2h; (2) 300 °C / 2226 h in vacuum	(1) 9.3×10^{-4} (2) 9.58×10^{-5}	(1) 3.04×10^{-5} (2) 1.92×10^{-4}	~5×10^{-5} at 300 °C	N.R.	[*24*]
W/Ni/Al and comparisons	6H and 4H	p	10^{19}	600 °C for 300 h	4×10^{-5}	Decreased slightly	——	N.A.	[*25*]
1) Ti 2) CrB_2	6H	p	1.3×10^{19}	1) 500 °C / 250 h 2) 300 °C / 1200 h	1) 2×10^{-5} 2) 3×10^{-5}	1) N.R. 2) unchanged	——	N.R.	[*26*]
W/Ni/Al/SiC	6H and 4H	P	10^{19}	400 °C / 1098 h	4×10^{-5}	Approx. unchanged	——	Tungsten carbide + Ni silicide	[*27*]
Al/TiW	6H	n	$>10^{19}$	400 °C / 560 h in air	5×10^{-4}	5×10^{-4}	——	Al oxide at surface	[*28*]
1)Au/TiC/ Ti_3SiC_2 2)Au/TiCN/ Ti_3SiC_2 3)Au/Pd/Ti_3SiC_2 4)Au/W/Ti_3SiC_2 5)Au/Ti/Ti_3SiC_2	6H	n	1×10^{18}	600 °C / 90 h in 93.5% Ar-6.5% H_2	N.R.	1) 10^4 X increase 2) 730 X increase 3) < 5 X increase 4 and 5) unmeasurable	——	N.R.	[*13*]
1)WSi_2 2)$WSi_{1.8}$	6H	n	$>10^{19}$	400 °C / 150 h in air	1)2.1×10^{-5} 2)7.2×10^{-5}	1)2–20 $\times 10^{-5}$ 2)1.1×10^{-4} (w/ Al encap.)	——	No interface reactions	[*10*]

3)$MoSi_{2.5}$					3)3.9 x 10^{-5}	3)1.7 x 10^{-5} (w/ Al encap.)		observed; oxides at surface	
Pt/TiN/Ti	6H	n	1.9 x 10^{19}	650 °C / 65 h in air	1.5 x 10^{-5}	No change	———	N.R.	[29]
1)TaC 2)Au/TaC 3)Pt/TaC 4)W/WC/TaC	6H	n	7.8 x 10^{18}	400 °C / 144 h in vacuum	1)8.04 x 10^{-5} 2)2.01 x 10^{-6} 3)1.39 x 10^{-5} 4)2.31 x 10^{-5}	1)8.38 x 10^{-5} 2)2.45 x 10^{-6} 3)1.63 x 10^{-5} 4)9.55 x 10^{-6}	1)N.R. 2)unmeasurable 3)7.50 x 10^{-7} @ 400 °C (on 2 x 10^{19} cm^{-3} sub.) 4)N.R.	No reactions observed between TaC/SiC	[8, 30]
Ti	3C	n	N_D N.R.; carr. conc. (by Hall) = 1 x 10^{16}	600 °C / 20 h in air	N.R.	15-25% increase	———	TiC (>600°C); Ti silicide (>700°C)	[31]
Pd	4H	p	5 x 10^{19}	500 °C / 100 h in N_2	5.5 x 10^{-5}	No change for contacts initially annealed at 700 °C for 5 min.	6.2 x 10^{-5} @ 450 °C (w/ Au overlayer)	Pd silicides and unreacted C	[32]
Al/$TiSi_x$	6H	n	5 x 10^{18}	———	7 x 10^{-6}	———	No change to 400 °C in vacuum	Ti_3SiC_2 and $TiSi_x$	[33]
Al/Si	4H	p	5 x 10^{19}	500 °C / 100 h in N_2	2.3 x 10^{-4}	No change	8.9 x 10^{-5} @ 450 °C in air	N.R.	[34]
TiC	4H	p	1–2 x 10^{19}	N.R.	1.9 x 10^{-5}	———	3.7 x 10^{-5} @ 300 °C	N.R.	[35]

W/WC/TaC	6H	n	7.8 x 10^{18} and 8.1 x 10^{18}	1) 600 °C / 1000 h in vacuum 2) 1000 °C / 1000 h in vacuum	6.6 x 10^{-5} (on 7.8 x 10^{18} cm^{-3}); 3.02 x 10^{-5} (on 8.1 x 10^{18} cm^{-3})	1) No change 2)7.51 x 10^{-4}	——	1) W_2C and TaC 2) W_2C and Ta oxides	*[17, 30]*
TiW	4H	n and p	1.3 x 10^{19} (n); 6 x 10^{18}and1.3 x10^{19} (p)	500 °C / 168 h in vacuum	1.2-1.7 x 10^{-4} (n); 4.0 x 10^{-5} (p)	No change (w/ Au overlayer)	~4 x 10^{-5} @ 300 °C (n)	N.R.	*[36]*
TiW	4H	n and p	1.1–1.3 x 10^{19} (n); 6 x 10^{18}, 1.3 x 10^{19} (p)	500 °C / 150 h + 600 °C / 150 h in vacuum	2–3 x 10^{-5} (n); 1.2 x 10^{-4} – 4.0 x 10^{-5} (p)	5–6 x 10^{-5} (n)	~1 x 10^{-5} @ 300 °C (n)	N.R.	*[12]*
1)Pt/Ti/TiW 2)Au/Ti/TiW	4H	n	1.1–1.3 x 10^{19}	500 °C / 150 h + 600 °C / 150 h in vacuum	1) ~6 x 10^{-5} 2) ~2 x 10^{-5}	1) ~6 x 10^{-5} 2) ~2 x 10^{-5}	——	N.R.	*[12]*
Pt/$TaSi_2$/Ti	6H and 4H	n	7 x 10^{18}	600 °C / 1000 h in air	1.68 x 10^{-4} (6H); ~4.7 x 10^{-4} (4H)	4.6 x 10^{-5} (100 h, 6H); ~3 x 10^{-5} (1000 h, 6H)	——	SiO_2, Pt silicide, Ti_5Si_3, Ti_xC_y	*[18]*
1) Pt(100nm) 2) Pt(100nm) / Si(10nm) 3) Pt(66nm) / Si(50nm)	6H	p	7.0 x 10^{18}	600 °C / 60 h in vacuum	1) 3 x 10^{-2} 2) 5 x 10^{-3} 3) 1 x 10^{-3}	1) unmeas. 2) 2 x 10^{-2} 3) 1 x 10^{-3}	——	1) and 2) Pt, Pt_2Si and PtSi 3) PtSi	*[14]*

From Table 1 it is seen that Au, Pt, Al and W have often been used as the top layers of multiple-layer contact schemes. The top layers are intended to prevent oxidation of the underlying layers and to connect to the outside world (typically via wire bonding to Au). However, the addition of these layers introduces new problems, which are dependent on the particular metal and underlying metallization scheme. For example, Au can melt and/or form a discontinuous layer due to its low melting point or its low eutectic temperature (363°C) with Si.[15] This low eutectic temperature was attributed to the formation of droplets in Au/Ti contacts after annealing at 600°C for 30 minutes.[16] In that case, much better results were obtained by substituting Pd for Au. While Al as an overlayer is said to produce a "passivating" oxide, the oxide is also electrically insulating, which results in a substantial increase in the resistance. In the case of TaC contacts, we found that W/WC overlayers served best as stable, passivation layers at 600°C in vacuum.[8,17] However, due to practical considerations Okojie *et al.*[18] emphasized the need to test contacts in air. Therefore, Pt was used as the overlayer in their study to optimize Pt/$TaSi_2$/Ti/SiC samples for stability at 600°C. Increasing the thickness of the $TaSi_2$ layer slowed the kinetics of the reactions between the layers, but reactions still occurred due to inherent instabilities in the system.

There is some evidence that the formation of carbon as a byproduct of a metal-silicide reaction is detrimental to the stability of contacts. In one study,[21] W/Ti/Ni/SiC and W/Mo/Cr/SiC metallization schemes were investigated to prevent the formation of "free" carbon. Although the contacts are reported to be stable after annealing at 650°C for 1000 h, it was not reported whether stability was measured in terms of contact resistance, phase formation or morphology.

Wire bonding (usually performed with Au) to the contacts can also be a problem. Although seemingly mutually exclusive, adhesion and interdiffusion between Au and annealed Ni contacts are reported challenges. Substitution of NiCr for Ni led to better adhesion and less interdiffusion between the contact layer and the Au cap layer.[37] This benefit was attributed to the reduction of free carbon formation via reaction with Cr.

To develop thermally stable contacts it is important to understand why and how contacts degrade under certain conditions. We have studied degradation mechanisms of TaC ohmic contacts with W/WC overlayers as a function of annealing temperature and time. Although the specific contact resistance (SCR) remained stable within the experimental error after annealing at 600°C for 1000 h, annealing at 1000°C resulted in a slight increase in the SCR after approximately 300 h and a substantial increase after 900 h. Auger electron spectroscopy and TEM showed no change in the TaC/SiC interface after 1000°C for 600 h; however, slight changes in the SCR and sheet resistance values were detectable, which is indicative of the high sensitivity of the electrical measurements. Slow increases in O concentration with annealing times > 600 h, as detected by SIMS, corresponded with increases in the SCR and sheet resistance values. The dramatic increase in O at the TaC/SiC interface after 900 and 1000 h also corresponded with the dramatic increases in SCR and sheet resistance values at those times. Extensive interdiffusion of the layers also occurred after 900 and 1000 h. Therefore, for annealing times less than 600 h, it was concluded that the small changes in the electrical properties were primarily due to O diffusion in the contacts. The dramatic increases in the contact and sheet resistance values for $t \geq 800$ h were believed to be caused by both the metallurgical reactions (i.e. interdiffusion, phase transformations and interfacial roughening) and the oxidation.

In general, high-temperature ohmic contact schemes must, at a minimum, meet the following criteria:

(1) The individual metal layers must not react with each other or with the semiconductor within the operating temperature range. Otherwise, if true thermodynamic stability among the layers does not exist, reactions within this temperature range must be sufficiently slow such that the resulting changes in the electrical properties, morphology, etc. are within the acceptable limits during the projected lifetime of the device;

(2) Oxidation of the contacts must be prevented;

(3) Contact resistances must stay within a certain range specified by the particular device parameters;

(4) Mechanical integrity, or adhesion, between the layers must be satisfactory;

(5) One must be able to connect the device to the outside world (e.g. via wire bonding with Au).

2.2. Thermal stability of rectifying contacts

Rectifying, or Schottky, contacts to SiC are of interest for a variety of devices, such as Schottky diodes and MESFETs. Because of the wider band gap, SiC Schottky diodes can operate at higher temperatures than Si Schottky diodes while maintaining higher current rectification ratios (> 10^6 at 300°C[38]) and higher blocking voltages (> 1000V at 300°C[39]).

Table 2 summarizes several recent studies of the effect of annealing or elevated operating temperature on the Schottky barrier heights (Φ_B). Schottky contacts on both n- and p-type 6H- and 4H-SiC typically have barrier heights ≥ 1.0 eV. It can be seen that W-based (W, WC, WN, TiW) and Ni contacts were frequently investigated as high-temperature Schottky contacts, whereas Pd has been frequently investigated for high-temperature gas sensors[40,41] because of its sensitivity to hydrogen.

Tungsten-based Schottky contacts are potentially useful at higher temperatures than nickel contacts because of the higher reaction temperatures of the former. Nickel contacts react with SiC at T > 600°C[39] and are known to become ohmic at T ~ 900°C. In comparison, W-based contacts may not react with SiC at temperatures up to 1200°C. The use of Ni is advantageous because it yields a high Schottky barrier height on n-type SiC, a factor that is important for low leakage currents at high temperatures. Secondly, processing is simplified if Ni is used for both the ohmic and Schottky metallizations. Because of its thermal stability and its demonstration as an ohmic contact to SiC, WC may also be a candidate for thermally stable ohmic and Schottky contacts to SiC.[46]

Long-term stability in an oxidizing ambient will likely be an issue for all of these contacts. We know of no studies that report the characteristics of Schottky diodes after extensive (> 1000 h) operation at elevated temperatures. Many of the same issues that were described for ohmic contacts, such as intermetallic reactions, diffusion barriers and oxidation, will also pertain to the thermal stability of Schottky contacts.

In one study[52] Re and $TaSi_2$, both of which are predicted to be stable with SiC, were investigated as Schottky contacts to 6H-SiC. Diffusion barriers of Ta-Si-N between the $TaSi_2$ and Au overlayers appeared to be successful for 1 h anneals at 700°C. The Re contacts showed an increase in SBH after a 2 h anneal at 700°C but did not change when annealed for an additional 2 h. Rhenium would appear to be a promising candidate for Schottky contacts to n-type SiC based on Re's relatively high work function and its predicted stability with SiC.[7] In contrast, several silicides and carbides have relatively low work functions[53] and are stable with SiC. In addition to their potential use as ohmic contacts, some of these compounds or alloys may make suitable Schottky contacts, particularly to p-type.

Because the Schottky barrier height is the most fundamental quantity used to characterize metal-semiconductor contacts, we have used this parameter as a focal point for summarizing the Schottky contact studies listed in Table 2. In most cases the calculated SBH either decreased or remained approximately constant as the measurement temperature was increased. At least part of the decrease is often attributed to the decrease in the SiC band gap with increasing temperature. Care must be taken in calculating and comparing the SBHs at high and low temperatures, as certain errors can be introduced. For example, high ideality factors, which are sometimes reported at lower temperatures, will likely result in underestimations of the SBH. Therefore, SBHs that were calculated in instances where high ideality factors were reported were omitted from Table 2. Additional errors may be introduced from uncertainties in parameter changes as a function of temperature (e.g. carrier concentration and effective mass).

Some investigators[50,51,54,55] have found evidence for two Schottky barrier height regions in diodes fabricated on 6H- and 4H-SiC. The low SBH regions in Ni diodes on 4H-SiC[50] were removed after annealing at 650°C. Various causes, such as doping variations, crystalline defects and surface contamination, have been blamed for the secondary barriers. However, the improvements reported after annealing indicate that the sources of the low SBH are located near the SiC surface. Removal of the low SBH regions results in a reduction in the reverse leakage current, an increase in the on/off ratio and an increase in the blocking voltage.[50] The low SBH regions tend to show effects at lower temperatures and voltages, where carriers are less likely to be promoted over the higher SBH region.

The results of reference 48 indicate that increasing the operating temperature yields more ideal Schottky contacts on p-type SiC. In that study, the ideality factors of $Ti_{0.58}W_{0.42}$ contacts on p-type 4H-SiC decreased from 3.11 to 1.16 as the temperature was increased from R.T. to 300°C, indicating that thermionic emission was only dominant at high temperatures. At the lower temperatures the current transport was attributed primarily to recombination. However, after annealing the contacts at 500°C for 30 minutes, the ideality factors were also low at R.T., indicating that thermionic emission was dominant throughout the temperature range. These results were attributed to an improved interface produced by annealing.

Table 2. Summary of results pertaining to the thermal stability of rectifying contacts to SiC. *Note: Multi-layered contacts are designated with slashes to separate the distinct layers; layers at the surface to the interface with SiC proceed from left to right. The metal on the top is listed first, and the SiC substrate is inferred last, e.g. Au/NiCr/SiC.

Metallization*	Substrate Polytype	Doping Type	Annealing Conditions	Φ_B (eV) before anneal	Φ_B (eV) after anneal	Φ_B (eV) at elevated temperature	Phases Formed	Ref
W	6H	n p	800 °C / 2 hrs in vac	1)0.87 2)1.80	1)0.79 2)1.57	1)0.84 @ 172 °C 2)1.67 @ 200 °C	None	[*42, 43*]
ITO (90% In_2O_3 – 10% SnO_2)	3-SiC/Si	p	No anneal; deposited at 200 °C	1.6	——	~1.3 at 200 °C	N.R.	[*20*]
Ti/Au/Pt/Ti	4H	n	500 °C / 100 h in N_2	1.17	——	No change @ 350 °C in air	N.R.	[*44*]
WN	4H	n	500 °C / 100 h in N_2	0.91	No change	No change @ 350 °C in air	W carbide and silicide @ 1200 °C / 4 min.	[*45*]
Ni	4H	n	600 °C / 5 min. and 900 °C / 5 min.	1.38 (I-V); 1.45 (C-V)	——	1.33 (I-V); 1.08 (C-V) @ 450 °C	None @ T ≤ 600 °C	[*38*]
Pd	6H	n	——	——	——	0.512 @ 500 °C in air; 0.014 @ 500 °C in 800 ppm H_2 in air	N.R.	[*41*]
WC	6H	1)n 2)p	Deposited @ 900 °C	——	1)0.79 2)1.81	1)0.89 @ 125 °C 2)1.91 @ 200 °C	No change	[*46*]

Pt	4H	n	200–1000 °C / 1 h	~1.7	~1.4 after 800 °C; ~1.0 after 1000 °C	———	Pt_3Si, Pt_2Si, PtSi and C (above 500 °C)	[*47*]
$Ti_{0.58}W_{0.42}$	4H	1) n 2) p	500 °C / 30 min. in vac	1)1.22 (I-V); 1.23 (C-V) 2)2.11 (C-V)	1)1.18 (I-V);1.19 (C-V) 2)1.91 (I-V);1.66 (C-V)	1)1.24 @ 300 °C (I-V) 2)2.01 @ 300 °C (I-V)	None	[*48*]
1)Ti 2)Ni 3)Au	4H	p	300 °C / 30 min. in vac	1)1.94 (I-V); 2.07 (C-V) 2)1.31 (I-V); 1.56 (C-V) 3)1.35 (I-V)1.49 (C-V)		1)1.97 (I-V) 2)1.23 (I-V) 3)1.05 (I-V)@ 300 °C	N.R.	[*49*]
Ni	6H	n	800 °C / 1 min. in Ar	Primary: 1.017; Secondary: 0.929	Primary: 1.232	———	N.R.	[*50*]
1)Ni 2)Pt 3)Co	6H	n and p	———	———	———	1)$\Phi_{Bn}^{I-V} = 1.00 +5.2x10^{-4}(T-300)$; $\Phi_{Bn}^{C-V} = 1.10 +6x 10^{-5}(T-300)$; $\Phi_{Bp}^{I-V} = 1.62 +4.5x10^{-4}(T-300)$; $\Phi_{Bp}^{C-V} = 2.18 -7x 10^{-5}(T-300)$ 2)$\Phi_{Bn}^{I-V} = 1.31$; $\Phi_{Bn}^{C-V} = 1.55 +5.6x 10^{-4}(T-300)$	N.R.	[*51*]

						3)$\Phi_{Bn}^{I-V} = 0.97 + 7.9x10^{-4}(T-300)$; $\Phi_{Bn}^{C-V} = 1.39 + 2.8x 10^{-4}(T-300)$		
Al/Ti	4H	n and p	——	——	——	$\Phi_{Bn}^{I-V} = 1.54 + 1.1x 10^{-3}(T-300)$; $\Phi_{Bn}^{C-V} = 1.47 - 2.1x10^{-3}(T-300)$	N.R.	[51]

Previous results[1, 56-58] show that the SBHs on 6H and 3C-SiC are strongly correlated with the work function of the metal contacts. The slopes of SBH vs. metal work function were in the range of 0.4–0.6, indicating only a partial pinning of the Fermi level in both 6H- and 3C-SiC. This parameter is an important characteristic of the semiconductor, because it gives an indication of the degree of control one has over the resulting SBH via the choice of metal. More recent studies by Itoh and Matsunami[59] and Lee *et al.*[49] also show strong correlations between the metal work functions and SBH's on 4H-SiC. The slopes on both n- and p-type 4H-SiC were similar to those reported for 6H- and 3C-SiC. In addition, the sum of Schottky barrier heights on n- and p-type material (Figure 2.3) was approximately equal to the band gap of 4H-SiC, which agrees with the Schottky-Mott model.

Raynaud *et al.*[51] also found correlations between the Schottky barrier heights and metal work functions on n- and p-type 4H- and 6H-SiC. They calculated the barrier height variations as a function of temperature and determined that they corresponded with the variations of the band gaps as a function of temperature. The I-V measurements were believed to yield underestimations of the barrier heights at low temperatures (< 300K) where high ideality factors were displayed. In this study, as in others, the effective masses of the carriers were assumed to be constant with temperature.

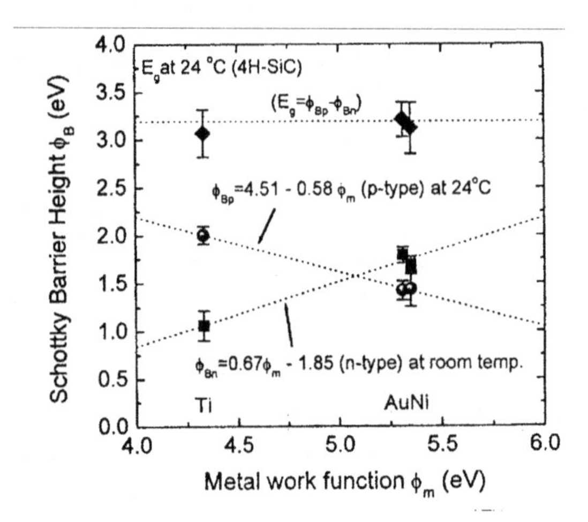

Figure 2.3. Schottky barrier heights of Ni, Au, and Ti on n- and p-type 4H-SiC as a function of metal work function. Reproduced from reference 49 with permission. The data for n-type SiC is from reference 59 and p-type SiC from reference 49.

3. Ohmic Contacts to p-type SiC

3.1. Introduction

The formation of reproducible, thermally stable and low-resistance ohmic contacts to p-type SiC remains a critical problem for reliable performance of certain SiC-based devices. Because of its large bandgap (3.0 eV for 6H-SiC and 3.2 eV for 4H-SiC) and high work function (4.8 eV for intrinsic SiC),[60] it is difficult to form ohmic contacts to p-type material by reducing the Schottky barrier height at the interface between the metal and the SiC. As shown in Figure 3.1, a significant SBH results at the metal/p-type SiC interface because the work functions of most transition metals, which are between 4 and 5.5 eV, are significantly less than the work function of p-type SiC. Therefore, reduction of the barrier width via high doping at the surface, instead of reduction of the barrier height, has been the preferred method for obtaining ohmic contacts to p-type SiC. As shown in Table 3, annealing the contacts at high temperatures (> 700°C) is also normally required. Other techniques involve selectively doping the surface layer of SiC during growth or by ion implantation.

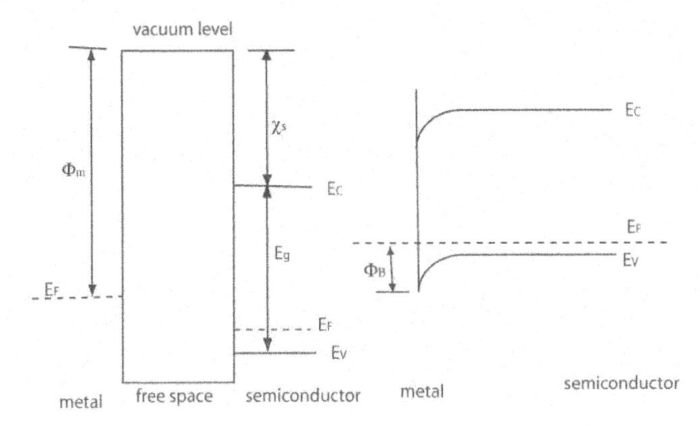

Figure 3.1. Energy band diagrams, as predicted from the Schottky-Mott relationship, before and after contact for a typical metal and p-type SiC. E_F = Fermi level, Φ_M – metal workfunction, E_C = conduction band minimum, E_V = valence band maximum, E_g = bandgap, χ_S = electron affinity and Φ_B = Schottky barrier height.

Table 3. Summary of results pertaining to ohmic contacts to p-type SiC. *Note: Multi-layered contacts are designated with slashes to separate the distinct layers; layers at the surface to the interface with SiC proceed from left to right. The metal on the top is listed first, and the SiC substrate is inferred last, e.g. Au/NiCr/SiC. N.R. = not reported.

Metallization	Poly-type	Doping Level (cm^{-3})	Implantation Conditions	Annealing Conditions	ρ_c (Ωcm^2)	Phases Formed	Morphology	Ref
Ti (250nm)/ Al (20nm)	6H	1.2×10^{15} to 1.4×10^{18}	Al at 50keV, Dose: 3.3×10^{3} cm^{-2} to 1.0×10^{15} cm^{-2}	20 – 1320 °C for 30min in pure Ar	1×10^{-2} for nonimplanted 5.4×10^{-4} for implanted	N.R.	N.R.	*[61]*
Al/Ti/Al	6H	$1 \times 10^{17} - 1 \times 10^{21}$	NA.	650 – 950 °C	High doping ohmic at 650C. Low doping ohmic at 800C	N.R.	N.R.	*[62]*
TiN (1000 A)	6H	1×10^{19}	FIB of Ga at 20 to 50keV; Dosage: 8×10^{14} to 5×10^{16} cm^{-2}	RT (amorphous), 600 °C to 800 °C	Low eV better contact Crystalline TiN better contact. 4.4×10^{-5} for TiN at 600C and 20keV Ga ions	N.R.	No interfacial reaction	*[63]*
Al	6H	1×10^{16}	a)Al, 4×10^{20} cm^{-3} b)Al + C, 4×10^{20} cm^{-3} both at 600C	950 °C for 5min	mid 1×10^{-2} mid 1×10^{-5}	N.R.	N.R.	*[64]*

a)Al(90wt%) – Ti(10wt%) b)Ti	6H	1.3 x 10^{19}	–	a)1000 °C, 2min b)800 °C, 1min	a)5 x 10^{-6} – 3 x 10^{-5} b)2 x 10^{-5} – 4 x 10^{-5}	N.R.	a)Severe int. reaction b)Smooth interface	[65]
Si/Co/SiC	6H	2 x 10^{19}	-	500 °C, 5h + 900 °C, 2h at 2 x 10^{-5} Pa	3.6 x 10^{-6} to 9.9 x 10^{-5}	$CoSi_2$	No interfacial reaction	[19]
a)Pt(800A)/Si (200A) b)Al(250A)/Ti (1500A)	4H	1 x 10^{19}	-	a)1100 °C for 3min b)1000 °C for	a)Low 1 x 10^{-4}	N.R.	Pt reacts with SiC; C moves to the surface	[66]
Os	6H	1.1 x 10^{19}/7.0 x 10^{18}	-	1020 °C for 1hr	6.8 x 10^{-4}	Silicides + C	Si and C rich surface layer	[67]
Pd	4H	a)3 x 10^{19} b)5 x 10^{19}	-	700 °C for~15min	a)5.6x 10^{-4} b)5.5 x 10^{-5}	Pd_2Si Pd_3Si	Hills and valleys on surface	[32]
W/Ni/Al Mb/Ni/Al Au/Ni/Al	6H 4H	1 x 10^{19}	-	850 °C for 2min	4 x 10^{-5}	AlNi, Ni_2Si	Carbon at interface	[25]
CrB_2 W_2B	6H	1.3 x 10^{19}	-	1100 °C for 2min and for 90min	1.9 x 10^{-5}(90min) 1 x 10^{-4}(2min)	N.R.	O removed by annealing	[68]
Pt- Si	6H	7 x 10^{18}	-	1100 °C for 5min	1 x 10^{-3}	PtSi, singlephase	Smooth interface	[14]
Al-Ti alloy	6H	7 x 10^{18}	-	1000 °C for 2min	1.5 x 10^{-4}	N.R.	Melting, spiking	[69], [70]
	4H	a)1 x 10^{18}	-	a)1100 °C for	a)2-3 x 10^{-4}	N.R.	N.R.	[71]

Ni(50nm)/Pd (10nm)/Ti(100nm) /Al(50nm)		b)1 x 10^{20}		1min b)1300 °C for 6 min	b)8 x 10^{-5}			
Al –2%Si-.15%Ti	4H	3-5 x 10^{19}	-	950 °C for 5min	9.6 x 10^{-5}	N.R.	N.R.	[72]
a)TiAl(%Al>77) b)TiAl(%Al <75)	4H	-	-	1000 °C	a)1 x 10^{-5} b) Not ohmic	a)Al_3Ti, Ti_3SiC_2, Al_4C_3 b)Al_3Ti, Ti_3SiC_2, Al_4C_3	a)Al agglomera-tion Rough morphology b) Smooth morphology	[73]
Co(10 nm)/Al (40 nm)	4H	9 x 10^{18}	-	900 °C for 5min	4 x 10^{-4}	N.R.	Surface roughness decreases with Al %	[74]
Al/Si(150nm total thickness)	4H	3-5 x 10^{19}	-	700 °C for 30min	2.3 x 10^{-4}	N.R.	Small spherical hills on top	[75]
a)Al+Si(150 nm total thickness) b)Pd	4H	5 x 10^{19}	-	a)700 °C for 20min b) 700 °C for 15min	a)3.8 x 10^{-5} b)5.5 x 10^{-5}	N.R.	N.R.	[76]
CrB_2	4H	1.3 x 10^{19}	-	1100 °C for 2min	8.2 x 10^{-5}	N.R.	Thin reaction region	[24]
Pd	4H	5 x 10^{19}	-	700 °C for 30min	5.5 x 10^{-5}	Pd_3Si, Pd_2Si, C	N.R.	[77]

Ti	4H	$>10^{20}$	-	-	3.4×10^{-4}	N.R.	Epitaxial	[*11*]
TiC	4H	$>10^{20}$	-	950 °C for 3min	2×10^{-5}	No interfacial reaction	Epitaxial, columnar grains	[*11, 12*]
TiW(30:70)	4H	1.3×10^{19}	-	980 °C for 1min	4×10^{-5}	No reaction	N.R.	[*12, 36*]

3.2. Al-based contacts

The traditional approach to forming ohmic contacts to p-type SiC is to anneal an Al-based contact (e.g. TiAl) on p+ SiC at temperatures between 900 and 1150°C.[78] However Al-based contacts are often considered to have morphological problems resulting from high processing and/or operating temperatures. These problems may be associated with the high tendency for oxidation, phase separation with low melting point phases, a low eutectic temperature between Al and Si, and/or a non-uniform interface with the SiC substrate. The ohmic behavior of the contacts is commonly presumed to be due to the diffusion of Al into the SiC, which enhances the doping of the surface layer. Correspondingly, the depletion region becomes thinner, and the conductance increases due to quantum-mechanical tunneling or thermionic field-emission through the barrier. This theory, however, has been repeatedly questioned[65, 69-71] by the revelation of pits and spikes at the interface, which suggests that Al may not actually dope the SiC surface, but may instead lead to enhanced field emission via the creation of many protrusions into the SiC material. These morphological features, i.e. spiking and pitting, pose a considerable challenge to the use of Al in the metallization scheme.

Some recent studies have focused on finding a suitable Al alloy that would lead to a lower contact resistance with better thermal stability and a more uniform morphology. In a study by Crofton *et al.*,[65] it was shown that during anneals of a 90/10 wt% Al-Ti alloy at 1000°C for 2 min in a vacuum furnace, significant amounts of Al were lost via evaporation. Other negative effects included spiking and pitting and inconsistency in the specific contact resistance values.

In a related study, Mohney *et al.*[69,70] tried several compositions of Al-Ti alloys as a contact layer. They concluded that the 70/30 wt% Al-Ti alloy showed the best results in terms of consistent contact resistances and reduced spiking. Compositions less than or equal to 60 wt% Al did not yield ohmic contacts. It was noted that the contacts with compositions > 60 wt% Al melted (in agreement with the Al-Ti phase diagram shown in Figure 3.2), and therefore the ohmic behavior may be associated with the formation of a liquid phase. These results have some correlation with other results[73] in which ohmic behavior was reported for a contact layer containing Al > 77 wt% and non-ohmic behavior for Al < 75 wt%. The contacts with higher compositions yielded a rough interface, whereas those with the lower compositions yielded a smooth interface. In another study[74] CoAl ohmic contacts resulted in better morphology and required a lower annealing temperature (800°C) than TiAl contacts.

Kassamakova *et al.*[75] reported the formation of Al/Si/ 4H-SiC ohmic contacts at temperatures as low as 700°C. Atomic force microscopy results indicated better morphology with no spiking into the SiC. This result was attributed to the addition of Si. Reductions in the specific contact resistances and improvements in the thermal stability were reportedly achieved via the addition of a minor amount of Ti[72] for reasons that are unclear.

Another problem with Al-based contacts is the high driving force for the oxidation of Al. The insulating oxide, which forms over time, increases the contact resistance to a potentially destructive level. Passivating layers, such as Ni and Pt[71] and W/Ni, Mo/Ni and Au/Ni,[19] have been investigated to prevent oxidation of the contacts and improve the thermal stability.

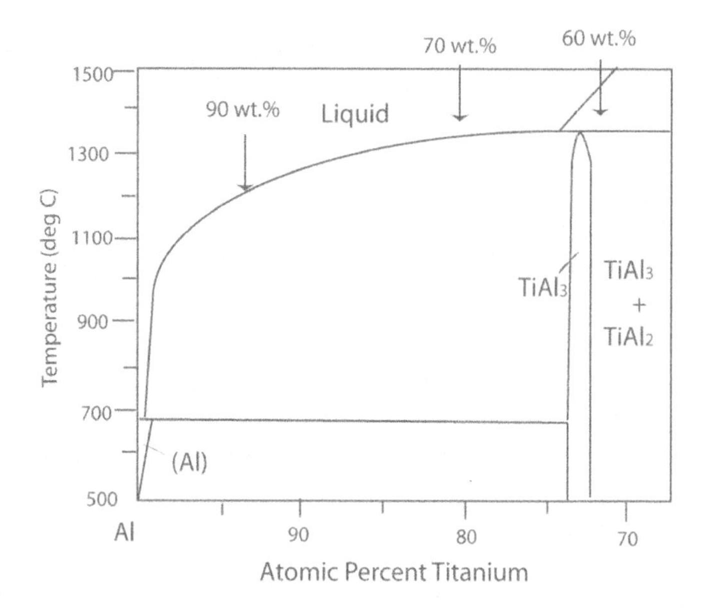

Figure 3.2. Portion of the Al-Ti phase diagram (after reference 15).

3.3. Al-Free Contacts

Because of the problems associated with Al, many researchers have investigated alternative metals. Elements such as Pt,[14,66] Pd,[32,76,77] Ti[11,79,36], Ni,[80] Co,[19] compounds such as TiC,[11,12] TiW,[36] CoSi,[19] and CrB_2,[24,68] and multilayer schemes including Pt-Si[14,66] and Co-Si-Ti[79] are described in the following text and/or in Table 3.

Refractory metal boride compounds are attractive materials for high-temperature electronic applications because of their reasonably high electrical conductivities and high melting points.[24,68] Rare earth borides can also act as diffusion sources for boron, which is a p-type dopant. Chromium boride, CrB_2, formed ohmic contacts on p+ (1.3×10^{19} cm^{-3}) 6H-SiC after annealing at 1100°C for a few minutes.[24,68] The interface, although rough, did not indicate any considerable amount of reaction between the CrB_2 film and the SiC substrate. A thin reaction region has been identified at the interface; however, its implications for the contact properties are not clear. Attempts to fabricate CrB_2 ohmic contacts on moderately doped p-type SiC (~1 x 10^{18} cm^{-3}) resulted in quasi-linear I-V characteristics and a high specific contact resistance. This implies that the use of CrB_2 contacts is limited to highly-doped p-type material (~ 10^{19} cm^{-3}). Attempts to wire bond to the Au overlayers on the annealed CrB_2 contacts indicate that an adhesion layer is necessary.

Refractory materials such as Co,[12,35,64] Pt[14,66] and Pd,[32,76,77] have also been investigated for ohmic contacts to p-type SiC. In some cases (e.g. Co, Pt and Al), a Si layer was deposited in addition to the elemental metal to improve the ohmic behavior. Lundberg *et al.*[19] formed $CoSi_2$ ohmic contacts after annealing a Si/Co bilayer structure. The Si layer prevented the formation of residual carbon as a result of a reaction between Co, which is a non-carbide former, and SiC. A larger number of groups appear to have had success with Pt-based ohmic contacts.[81] The advantages of Pt for p-type contacts include its high work function, its high melting point, and its oxidation resistance. Papanicolaou *et al.*[66] formed ohmic contacts on p-type SiC with a Si/Pt bilayer structure by annealing at 1100°C for 3 minutes. The authors point to the correspondence, and possible association, between the onset of ohmic behavior and evidence for melting at this temperature. In this study some out-diffusion of C was still observed, indicating that insufficient Si was present for complete reaction with the Pt film.

To investigate the role of Si in non-carbide-forming contacts, Jang *et al.*[14] compared Pt contacts with and without Si interlayers on p-type SiC. The Pt contacts were ohmic on moderately-doped (7 x 10^{18} cm^{-3}) p-type 6H-SiC after annealing at 1100°C for 5 minutes. It was found that in all cases under the range of conditions studied (e.g. Si thickness and deposition temperature), the incorporation of a Si layer reduced the contact resistance relative to Pt contacts that did not contain a Si layer. Moreover, independent reductions in the contact resistance were achieved via the use of B-doped Si, a higher deposition temperature (500°C) and the design of the Pt/Si layer thickness in a 1:1 atomic ratio. The latter condition yielded single-phase PtSi contacts (Figure 3.3) with a smooth interface and no indication of reaction with the SiC (Figure 3.4). The Si layer prevented the consumption of Si from the SiC and hence the formation of free C. The electrical characteristics of these contacts were stable after annealing at 600°C for 60 h, results that are in contrast to those observed for pure Pt contacts and for contacts containing a higher Pt:Si ratio.

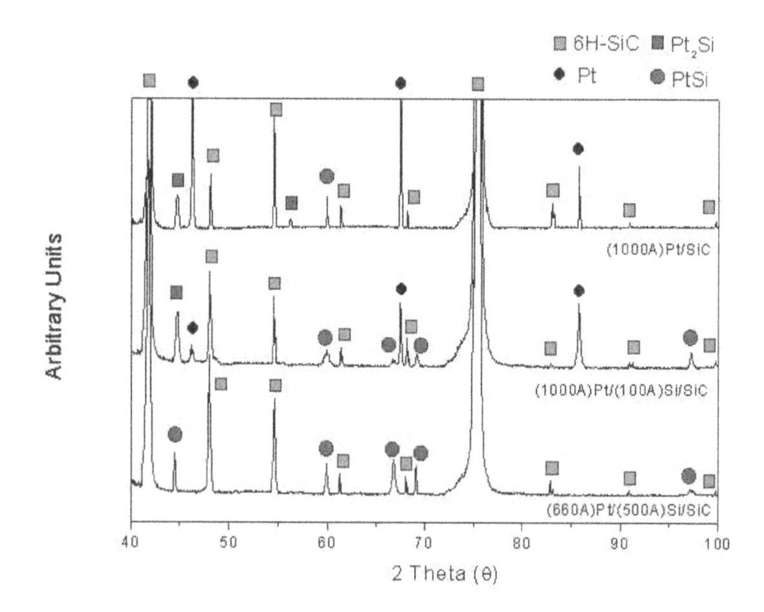

Figure 3.3. X-ray diffraction spectra for (a) Pt (1000 Å) / SiC, (b) Pt (1000 Å) / Si (100 Å) / SiC and (c) Pt (660 Å) / Si (500 Å) / SiC. The samples were annealed at 1100°C for min. (From reference 14).

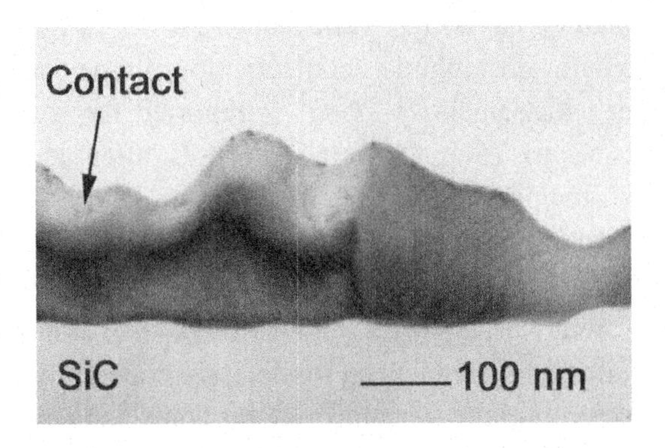

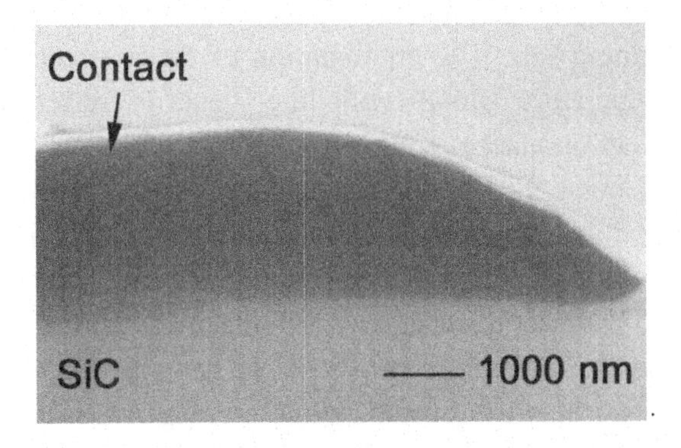

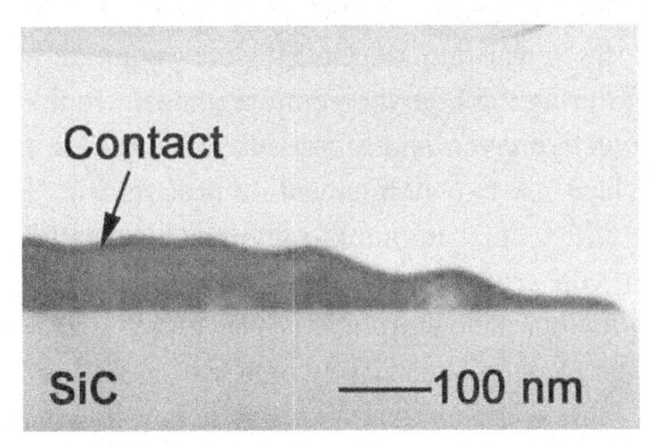

Figure 3.4. Cross-section TEM images of (a) Pt (1000 Å) / SiC, (b) Pt (1000 Å) / Si (100 Å) / SiC and (c) Pt (660 Å) / Si (500 Å) / SiC. The samples were annealed at 1100°C for 5 min. (From reference 14).

Palladium, like Pt, has a high work function (5.12 eV) and is therefore a good candidate for ohmic contacts to p-type wide bandgap semiconductors. Kassamakova *et al.*[32,77] reported the formation of Pd ohmic contacts to p+ (5×10^{19} cm^{-3}) 4H-SiC after annealing at the relatively low temperature of 700°C. While the high doping concentration of the substrate is expected to assist in the formation of the ohmic contacts, the lower reaction temperature of Pd with SiC may allow for the ohmic behavior at lower temperatures than other comparable contacts. A study of Pd contacts on moderately doped p-type SiC would be helpful to determine the versatility of Pd contacts to p-type SiC and compare it with other contacts, such as Pt. Because Pd is chemically similar to Pt, it may also be advantageous to incorporate a Si interlayer in contacts containing Pd. The combination of these effects, i.e. the low formation temperature along with the benefits observed from Si interlayers in Pt contacts, could result in a practical and versatile metallization scheme for low resistance ohmic contacts to p-type SiC.

Titanium, which has been widely used in metallization in silicon devices, has also been investigated for contacts to p-type SiC. In one study[65] pure Ti contacts resulted in low contact resistances ($2–4 \times 10^{-5}$ Ω cm^2) on p-type (1.3×10^{19} cm^{-3}) 6H-SiC after annealing at 800°C for 1 minute. The surface, as revealed after etching away the metal, was smooth, indicating a planar Ti/SiC interface. In another study[79] passivating layers consisting of Co/Si were required because of the oxidation of Ti during the high temperature anneal. In this case, Co was used as the protective layer, and Si was used to prevent the diffusion of Co to the interface. A two-step annealing process was developed (500 °C for 5 min + 850°C for 1 minute) to give the best contact resistance (4×10^{-4} ohm-cm^2) on a moderately-doped sample (3.9×10^{18} cm^{-3}). No information on the interface morphology was given.

An alternative to pure Ti is TiW, which also has a low resistance and can be etched with wet chemicals but has better resistance to reaction with SiC. Hence, it is a potential candidate for a contact to SiC at high operation temperatures. Lee *et al.*[12,36] sputtered TiW (30:70 wt ratio) contacts on highly- (1.3×10^{19} cm^{-3}) and moderately-doped (6×10^{18} cm^{-3}) 4H-SiC and annealed them at 980°C. The electrical properties of the TiW ohmic contacts did not change after annealing at 500°C for 168 hours in vacuum when covered with Au or with Pt. No phases were believed to form during the annealing process, and the surface morphology remained smooth.

Transition metal carbides are also of interest in general, because many of them are relatively conductive and inert with respect to reactions with SiC at elevated temperatures. Titanium carbide is of particular interest because it can grow epitaxially on SiC; epitaxial TiC ohmic contacts on p-type SiC have been reported.[11,12] Titanium and C_{60} were co-evaporated at a temperature of ~ 500°C. Titanium carbide ohmic contacts have the advantage of epitaxy and a smooth interface without any reactions.

4. Ohmic Contacts using Nickel

4.1. Introduction

Nickel has been the most commonly used metal for ohmic contacts to n-type SiC due to the fact that specific contact resistance (SCR) values < 1×10^{-5} ohm-cm^2 [82-84] are reproducibly obtained. In addition, annealed Ni contacts are reportedly stable up to temperatures of 600°C.[82] To produce ohmic contacts on 6H- or 4H-SiC, the Ni must be annealed at temperatures above 900°C (see Section 3.3). Below this temperature, Ni forms a Schottky contact with a high barrier height.

Crofton *et al.*[82] reported SCR values of less than 5×10^{-6} ohm-cm^2 for Ni contacts that were annealed at 950°C for 2 min on 6H-SiC with a doping level of 7×10^{18} cm^{-3}. The annealed metal film transformed to a nickel silicide, which contained C as a second phase. A summary of recent results pertaining to Ni ohmic contacts to n-type SiC is given in Table 4.

4.2. Thermodynamics

Figure 4.1 shows an empirical phase diagram for the ternary Ni-Si-C system at 900°C.[7] Since Ni does not have a tie line with SiC, it is not in thermodynamic equilibrium with SiC. Therefore, given sufficient time, Ni and SiC will react. Although Ni does not form any carbides, several

Ni silicides exist. Four of the silicides (Ni_2Si, Ni_3Si_2, NiSi, and $NiSi_2$) are stable with SiC. Because Ni_2Si has a tie line with both C and SiC, one would expect Ni_2Si and C to be the final products resulting from a reaction between Ni and SiC at 900°C.

Empirical results agree with this prediction. For example, a diffusion couple consisting of Ni and SiC annealed at 900°C for 24 h yielded a periodic banded structure of Ni_2Si and graphite.[103] A similar structure was also produced after depositing Ni films with a nominal thickness of 2.5 monolayers on a H_2- etched 6H-SiC substrate heated to 600°C.[104] Heating to 1000°C produced islands, one type of which consisted of Ni and C in alternating sheets and the other type of which consisted of Ni, Si and C.

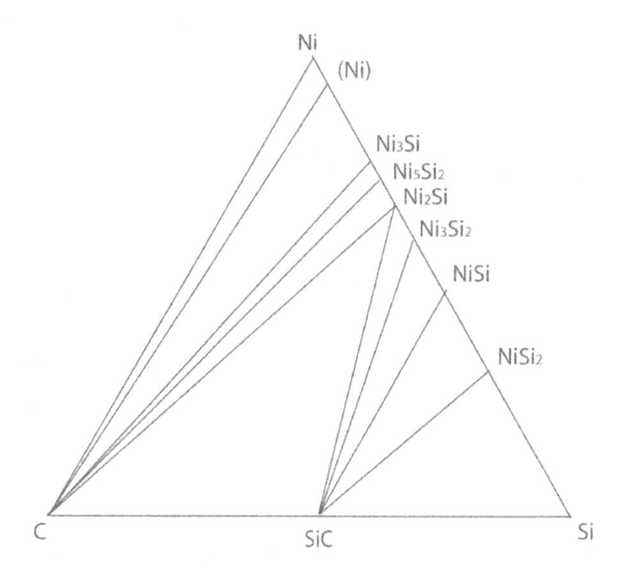

Figure 4.1. The approximate Ni-Si-C phase diagram at 900°C. After references 7 and 102.

4.3. Phase formation sequence

The phases reported in annealed Ni/SiC samples usually agree with the thermodynamic prediction. Several authors report that after a short anneal at temperatures above 900°C, Ni_2Si and C form. Some authors,[86,93,98,105] however, report NiSi as the final phase. Han *et al.*[98,105] found that the Si content in the film increased with increasing annealing temperature, hence resulting in the formation of NiSi (Ni:Si = 1) from Ni_2Si (Ni:Si = 2).

Studies have shown that Ni begins to react with SiC at temperatures as low as 500°C.[97,106] At the early stages of reaction $Ni_{31}Si_{12}$ is reportedly formed at the interface.[98, 101] Between temperatures of 700 and 800°C, Ni_2Si forms. The concentration of Ni_2Si in the film increases with increasing temperature. At temperatures above 900°C the $Ni_{31}Si_{12}$ phase disappears, and the predominant phases are Ni_2Si within the film and C at the surface. The effect of annealing time and film thickness on the microstructure and on the electrical properties is not evident from the information available. Longer annealing times and thinner samples should result in phases that approach or achieve equilibrium. Similarly, a thinner sample would take less time to complete the reaction. The sequence of phase formation is summarized in Figure 4.2.

$$Ni + SiC \xrightarrow{600°C} Ni_{31}Si_{12} + C + Si \xrightarrow{700°C}$$
$$Ni_{31}Si_{12} + Ni_3Si + C + SiC \xrightarrow{900°C} Ni_2Si + C + SiC$$
$$\xrightarrow{>1000°C} NiSi + C + SiC$$

Figure 4.2. Diagram depicting the sequence of phase formation in annealed Ni/SiC.

4.4. Issues

Post-deposition annealing at temperatures between 900 and 1050°C (the temperature range required to form ohmic contacts) causes the Ni to react with SiC to form nickel silicides.[82,93,96,107,108] Consequently, undesirable effects such as broadening of the metal-SiC interface, interface and/or surface roughening, formation of Kirkendall voids, and carbon segregation at the interface and throughout the metal layer may occur.[91] A representative AES depth profile of the reaction zone of a Ni contact on SiC annealed at 1050°C shows broadening of the interface and the segregation of carbon.[88] These features inhibit long-term reliability and ultimately cause device failure via contact degradation and/or wire bond failure after exposure to high powers or high temperatures. The voids at the interface cause internal stress, which can result in delamination of the contact layer. A rough surface morphology makes wire bonding difficult. The effects of carbon segregation are not well understood. Redistribution of the carbon during operation at high temperatures could result in electrical instability. Thus, some research efforts are focusing on obtaining contact layers that are free of carbon inclusions. At least one study[109] indicates that controlled deposition of C in Ni/C/SiC contacts may have beneficial effects when compared to the uncontrolled formation of C in annealed Ni/SiC contacts. The role of carbon in ohmic contacts is unclear and worthy of further investigation. In another approach, low-resistance contacts were fabricated on 3C-SiC epilayers via deposition of an amorphous Si layer at 550°C followed by the deposition of Ni and a low temperature anneal at 300°C.[121]

Table 4. Summary of results pertaining to Ni ohmic contacts to n-type SiC. *Note: Multi-layered contacts are designated with slashes to separate the distinct layers; layers at the surface to the interface with SiC proceed from left to right. The metal on the top is listed first, and the SiC substrate is inferred last, e.g. Au/NiCr/SiC. N.R. = not reported.

Metallization (thickness)	Substrate Polytype	Epilayer Doping (cm^{-3})	Annealing Conditions	ρ_c (ohm cm^2)	Morphology and Phase Formation	Thermal Stability	Ref
Ni(N.R)	6H	7 x 10^{18} – 9 x 10^{18}	950 °C for 2min in 10^{-6} torr vacuum	<5 x 10^{-6}	Ni_2Si with intermixed C and O throughout the layer	At 650 °C for 329 h: slight increase in ρ_c	[82]
Ni(200 nm)	6H	4.5 x 10^{20}	1000 °C for 5min in Ar	<1 x 10^{-6}	Rough surface, phases formed N.R.	N.R.	[83]
W(~10nm) /$Ni_{90}Ti_{10}$ (100nm)	6H (0001 face)	~1 x 10^{18}	800 °C in ~1 x 10^{-7} torr vacuum	N.R	Ni_2Si(~120nm) /TiC(~60nm) /Ni_2Si+C(~20nm) /SiC	N.R.	[85]
Ni(90nm) /Al(30nm)	6H	1 x 10^{18}	a)as-deposited b)1020 °C for 5min/1h in N_2:H_2(99:1)	a)4.4 x 10^{-4} b)1.2 x 10^{-4} /2.6 x 10^{-4}	Ternary compounds formed. NiSi gives ohmic behavior	At 1020°C for 20 h increases ρ_c to 7 x 10^{-4}	[86]
Ni(100nm)	6H	1x 10^{18}	1020 °C for a)5min b)1h in N_2:H_2(99:1)	a)2.1x 10^{-4} b)3.4 x 10^{-4}	C at subsurface region; Formation of NiSi	N.R.	[87]
Ni	6H a)(0001) b)(000)	2-5 x 10^{18}	1000 °C, Vacuum	8-9 x 10^{-5}	C enriched layer at interface; Ni_2Si with C throughout	N.R.	[88]

a)Ni(100nm) b)Al(20nm)/ Ni(75nm)/ Al(7nm)	4H$(11\bar{2}0)$	1 x 10^{19}	1000 °C, 5min in N_2	a)6 x 10^{-5} b)1.8 x 10^{-5}	Three regions: 1)Al_2O_3 2)Al_4C_3+Ni_xSi+SiOx 3)C rich, voids	N.R.	[*89*]
Ni(500nm)	3C	~1 x 10^{17}(carrier conc.)	a)as-deposited b)500 °C in Ar+8%H_2	a)5 x 10^{-4}(RT) b)5 x 10^{-5}	Reaction at 500C, Ni completely consumed at 900C. Reaction Products unidentified	N.R.	[*90*]
Ni-Cr(80:20 wt%) a)50nm b)200nm	4H	4.8 x 10^{17}	1100 °C for 3min in 1 x 10^{-6} torr,	~1.5 x 10^{-4}	a)microscopic islands; thin C on surface b)thick C on surface	N.R.	[*23*]
a)Ni(150nm) b)Ni-Si multilayers (150nm)	i)6H ii)4H	i)(1-1.8) x 10^{18} ii)1 x 10^{19}	950 °C for 10min in N_2	a)2.8 x 10^{-6} b)1.2-2.7 x 10^{-5}	a)C throughout, large reaction with SiC b)Little C at the interface. No C in the film Ni_2Si formed in both cases	N.R.	[*77*]
Pt(100nm)/ Ti(25nm)/ WSi(80nm)/ Ni(40nm)	4H	8 x 10^{18}	1000 °C for 30 s in RTA furnace	N.R	Smooth Surface; Ni_2Si phase at the interface; Residual C consumed by WSi and Ti	N.R.	[*91*]

Ni	4H	9.5 x 10^{18}	950 °C for 10min in N_2 + 10%H_2	N.R	Reaction with substrate; C in the film	N.R.	[*92*]
Ni(100nm)	4H	4.2 x 10^{15}	1000 °C for 1min in RTA furnace	2.8 x 10^{-3}	C at surface; NiSi phase at 950 C and Ni_2Si at 800 C	N.R.	[*93*]
Ni_2Si(400nm)	4H	2x 10^{18}	a)700 °C b)950 °C for 30 s in RTA furnace	N.R	No C; No voids; Sharp smooth interface	N.R.	[*94*]
a)Ni(30nm)/ Si(90nm) b)Ni	6H	1 x10^{18}	i)900 °C for 10min ii)1100 °C for 15min in Ar + 5% H_2	a) 3.9 x 10^{-3}	a)$NiSi_2$; No C; SiO_2 at surface & interface; Sharp smooth interface b)Ni:Si(2:1), Rough interface	N.R.	[*95*]
Ni(25nm)/Si (92.5nm)	6H	a)4.1 x 10^{18} b)2.5 x10^{19}	900 °C for 10min in Ar + 5%H_2	a) 3.8x 10^{-4} b) 3.6 x 10^{-6}	$NiSi_2$; No C; Smooth Sharp Intefaace	N.R.	[*84*]
Pt(100nm)/Ti (25nm)/WSi (80nm)/Ni (40nm)	4H	8 x 10^{18}	a)950 °C b)1000 °C for 30s in N_2 atmosphere	N.R	Smooth Surface; Ni_2Si phase at the interface; Residual C consumed by WSi and Ti	(b) remains stable after 100h at 650 °C, unlike (a)	[*96*]

Ni(200nm)	6H	1 x 10^{18}(carrier conc.)	1000 °C for 5min in Ar	3 x10^{-3}	Ni_2Si is the main phase; NiSi and $NiSi_2$ also form; C at the surface and in the film	N.R.	[*97*]
Ni(100nm)	4H	4.2 x 10^{15}	a)800 °C b)1000 °C	2.8 x 10^{-3}	a)Ni_2Si-$Ni_{31}Si_{12}$; C band near interface b)NiSi-Ni_2Si; C at the surface and in the film	N.R.	[*98*]
Ni(100nm)	6H	7.4 x 10^{18}	950 °C in N_2	3.6 x 10^{-5}	Ni_2Si; Rough interface; C uniformly distributed in film	N.R.	[*99*]
Ni(400nm)	6H	4 x 10^{17}	1000 °C for 5min a)Ar b)N_2	a) not ohmic b) ohmic	a)larger grain size; rough surface b)smaller grain size; smooth interface	N.R.	[*100*]
Ni(110nm)	4H	7 x10^{18}	To 1000 °C in RTA	N.R	$Ni_{23}Si_2$ +$Ni_{31}Si_{12}$ -> $Ni_{31}Si_{12}$ ->$Ni_{31}Si_{12}$ +Ni_2Si -> Ni_2Si; C in the film	N.R.	[*101*]
Ni(30nm)/Si (100nm)	3C	< 10^{18}	Si dep. at 550 °C; Ni/Si anneal at 300 °C / 12 h	5.6x10^{-5} to 3.0x10^{-3}	$NiSi_2$	N.R.	[121]

4.5. Mechanisms

Several mechanisms have been suggested for the ohmic behavior of Ni contacts to SiC. The formation of Ni_2Si at the metal-SiC interface begins at 600°C[87,88,97-99,105] and is complete at 900°C,[97] as indicated in Figure 4.2. Because ohmic behaviour is not observed after annealing at temperatures below 900°C, it has been proposed that the formation of nickel silicide is insufficient to ensure ohmic behaviour; instead, the completion of the reaction to form Ni_2Si as the primary phase is believed to play a dominant role.[97]

We further propose that the ohmic behavior may be associated with the formation of vacancies in the SiC substrate during the high-temperature annealing step. Carbon vacancies may act as donors, resulting in an increase in the net concentration of electrons under the contact. This phenomenon would result in a reduced depletion layer width and a consequent reduction of the contact resistance.[93,98,105] This reasoning is further evidenced by the observation that unannealed, nickel silicide layers do not exhibit ohmic behaviour on heavily-doped SiC epitaxial material,[94] which indicates that the annealing step itself plays an important role in the formation of ohmic contacts.

The formation of ohmic contacts of Ni/Si multilayers and of as-deposited $NiSi_2$, cases in which the reaction between the metal layer and the substrate is reduced, indicates that other mechanisms may also contribute to the ohmic behavior. The mechanisms may include a combination of various point defects and reduced barrier heights, although systematic studies are needed in order to make further conclusions regarding the mechanism(s) responsible for the ohmic behaviour.

4.6. Alloying

The undesired segregation of C can be reduced or eliminated by introducing a carbide-forming element (e.g. Si) in the metallization scheme. In some cases reaction between the metal layer and the substrate may be completely eliminated.[84,95,110,111]

Incorporation of Al in Al/Ni/Al contacts reportedly resulted in lower contact resistance values than Ni contacts that did not contain Al.[89] The first Al layer was believed to prevent the formation of voids and to reduce the oxide layer initially present on the SiC surface. However, some voids were still observed. After annealing at 1000°C for 5 minutes, the contacts consisted of three distinct layers. The upper layer of Al_2O_3 was believed to serve as a protective layer and therefore improved the thermal stability.[86] The C was present in the middle layer in the form of Al_4C_3. The ohmic behavior was attributed to the NiSi phase, which was located at the interface.

Ni_2Si films with little C at the interface were formed by annealing multilayers of Ni and Si at 950°C for 10 minutes. The Si layer reduces the consumption of the substrate, and hence the reaction zone is much thinner in comparison to annealed Ni/SiC structures. Ohmic contacts with specific contact resistance values between 1.2 and 2.7 x 10^{-5} ohm-cm^2 were reported.

The direct deposition of Ni_2Si contacts is also an option. Pulsed laser deposition of Ni_2Si films followed by an anneal at 950°C yielded ohmic contacts with an abrupt, void-free interface and a smooth surface morphology.[94] However, the specific contact resistance values were not reported.

Others have formed $NiSi_2$ contacts devoid of C by depositing Ni and Si in the required ratio and annealing at 900°C.[84,95,110,111] The contacts were ohmic and showed an abrupt interface. Although the reported contact resistance (3.8 x 10^{-4} ohm-cm^2) is relatively high (e.g. 5 x 10^{-6} ohm-cm^2 was reported for annealed Ni contacts on a moderately doped SiC substrate),[82] note that $NiSi_2$ is a thermodynamically stable phase with SiC (see Figure 4.1) and therefore holds promise for high-temperature applications. In comparison, on 3C-SiC, $NiSi_2$ ohmic contacts with ρ_c = 5.6 x 10^{-5} – 3.0 x 10^{-3} Ω cm^2 were demonstrated after deposition and annealing temperatures of only 550 and 300°C, respectively.

Figure 4.3 shows a comparison of the RBS spectra taken from the Ni/Si/SiC and the Ni/SiC samples. The difference in the abruptness of the interface is apparent. After annealing the Ni/Si/SiC sample at 900°C for 10 min, the deposited Ni film completely reacted with the Si film, but no reaction with the substrate was apparent. The estimated composition of Ni:Si = 1:2 is consistent with the ratio of Ni and Si atoms in the deposited films. Thus, it appears that by controlling the ratios of Ni and Si, reaction with the SiC substrate can be avoided. These results are similar to those reported for Pt/Si contacts to p-type SiC.[14]

To overcome problems with wire bonding to Ni contacts, nichrome (80:20 wt% Ni/Cr) ohmic contacts were used by Luckowski *et al.*[23] The electrical characteristics were found to be similar to those of Ni contacts. One disadvantage is that the nichrome contacts also result in the formation of unreacted C, too much of which can interfere with successful wire bonding. The addition of gold cap layers significantly improved the ability to wire bond. The composite Au/NiCr/SiC contacts remained stable after 2500 h at 300°C.

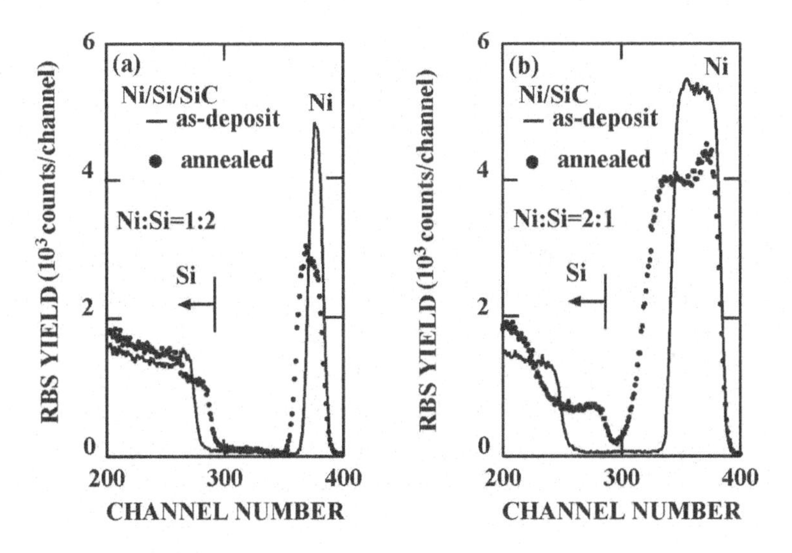

Figure 4.3. Comparison of the RBS spectra for (a) Ni/Si/SiC and (b) Ni/SiC samples before and after annealing. The annealing temperatures for the Ni/Si/SiC and the Ni/SiC samples are 900 and 1000°C, respectively (from reference 95). Reproduced with permission.

5. Effects of Defects on Schottky Contacts

Although the intrinsic properties of SiC allow for the fabrication of Schottky diodes with high blocking voltages with smaller drift regions and therefore smaller devices than Si, the main technological challenge to the realization of large-area SiC power devices is the identification and reduction of electrically-active defects in the SiC crystals. As such, in this section we will summarize evidence for the effects of micropipes, closed-core screw dislocations, stacking faults and reactive-ion-etch-induced defects on the properties of SiC Schottky diodes.

In the early to mid-1990s, micropipes were considered the primary culprit that limited device performance. Micropipes were shown to consistently cause premature breakdown in 4H- and 6H-SiC p-n diodes.[112]

Now that commercial suppliers have been able to substantially reduce the micropipe densities, electrical degradation due to other crystal defects have come under consideration. For example, in addition to the obvious effect of micropipes on limiting the breakdown voltage (BV) of SiC Schottky diodes, Wahab *et al.*[113] reported an inverse correlation between the number of closed-core screw dislocations and the breakdown voltage (Figure 5.1), in agreement with similar results reported for SiC p-n diodes.[114] To achieve BV's > 1.5kV in the Schottky diodes, a dislocation density < 2000 cm^{-2} was required. This phenomenon was attributed to an enhancement of the electric field around the dislocations. Others[115] have found a correlation between the locations of closed-core screw dislocations, identified by synchrotron white-beam x-ray topography (SWBXT) images, and dark spots apparent in electron-beam-induced current (EBIC) images of SiC Schottky diodes. The authors concluded that the screw dislocations reduced the diffusion length of carriers in the SiC by ~30%. In contrast, another study[116] involving EBIC and SWBXT characterizations indicates that the ideality factors and barrier heights correlated with the number of EBIC dark spots but not necessarily with the screw dislocation density. These results suggest that further studies are needed to elucidate the true nature of defects that affect the electrical properties (forward and reverse) of SiC Schottky diodes.

Inhomogeneities (e.g. defects or second phases) have long been known to affect the electrical characteristics of Schottky contacts. For example, Tung[117-119] attributes commonly-observed differences between I-V and C-V characterizations of Schottky diodes to inhomogeneous Schottky barriers within the contact region. Similarly, excess currents at low bias voltages in some SiC diodes may be attributed to low Schottky barrier height patches within a predominant, high Schottky barrier height (SBH) 'phase'.[55] Although the low SBH phase may cover only a very small fraction of the contact area, it can still significantly affect both the forward and reverse current-voltage characteristics. In the study by Skromme *et al.*[55] the average barrier heights calculated from C-V measurements were significantly higher than the values calculated from I-V measurements, especially for diodes with high ideality factors. Whereas the barrier height values determined from I-V measurements decreased with increasing ideality factor, the barrier height values from C-V measurements were independent of n (see Figure 5.2). These results agree quite well with those reported in reference 54.

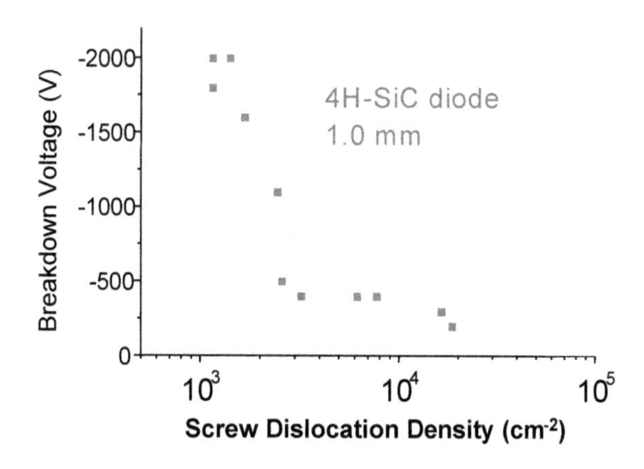

Figure 5.1. Plot of the breakdown voltage measured as a function of the screw dislocation density in 4H-SiC Schottky diodes (from reference 113). Reprinted with permission.

Processing-induced defects can also degrade the electrical properties of Schottky contacts. For example, reactive ion etching (RIE) is used to pattern SiC-based devices because of the lack of suitable wet etchants. The effects of RIE with CHF_3/O_2 on the electrical properties of 4H-SiC Schottky diodes were investigated.[120] The "ideal" barrier heights were determined by extrapolating plots of ideality factor vs. barrier height to n = 1.0. These values agreed quite well with the barrier heights calculated from C-V measurements, as predicted by Figure 5.2. The authors report that the barrier heights on etched SiC were much less sensitive to the metal work function, and larger ideality factors were displayed. The slope of barrier height vs. work function is 0.65 eV for unetched SiC and 0.11 eV for etched SiC, indicating strong Fermi level pinning in the etched SiC.

Stacking faults are additional defects that can form during the growth, processing or operation of SiC devices. The growth of stacking faults[122] during the operation of SiC bipolar devices resulted in the degradation of the forward bias characteristics.[123,124] Okojie *et al.*[125] reported the appearance of stacking faults in n+ 4H-SiC following thermal oxidation. Skromme *et al.*[126] reported morphological and structural degradation of n+ N-doped (3 x 10^{19} cm^{-3}) 4H-SiC substrates resulting from thermal oxidation at 1150°C. The barrier heights of Ti, Ni and Pt Schottky diodes in this material were reduced by 0.47 eV relative to lower-doped (non-degraded) material, irrespective of the metal contact. The structural defects were later identified by high resolution TEM as double stacking fault inclusions, which have the 3C-SiC stacking layer sequence.[116] The results support their theory that the reduced barrier heights are due to an increase in the electron affinity associated with the 3C-SiC layer.

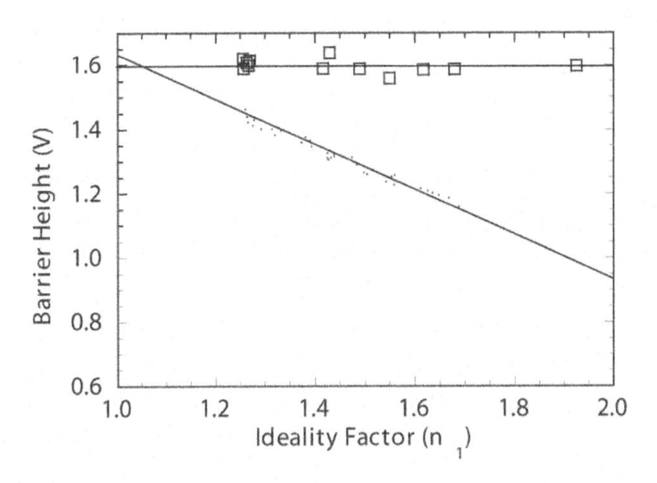

Figure 5.2. Barrier heights of 4H-SiC Schottky diodes calculated from I-V (dots) and C-V (squares) measurements as a function of the ideality factor (from reference 55). Reproduced with permission.

The $(11\bar{2}0)$ orientation has also been of interest because of the reported increase in the channel mobility in $(11\bar{2}0)$ MOSFETs. However, Schottky diodes on $[11\bar{2}0]$ epilayers had larger leakage currents and lower blocking voltages than diodes on [0001] epilayers.

These results were attributed to stacking faults that form during growth of the $[11\bar{2}0]$ epilayers. Based on the forward I-V characteristics, the authors concluded that the stacking faults reduced the Schottky barrier height. They plotted the barrier heights vs. the number of stacking faults. However, the ideality factors are not reported. It is therefore possible that the barrier heights are underestimated for the poorer diodes. In general, the evidence indicates that stacking faults, in addition to other types of defects, have significant detrimental effects on the properties of Schottky diodes.

6. Summary

The intrinsic properties of SiC offer numerous advantages over Si for use in a variety of extreme conditions. One of the primary applications considered for the commercial use of SiC is high-temperature electronics. However, interfaces between dissimilar materials at the discrete device level as well as at the packaging and interconnection levels often interact at relatively low temperatures and therefore cause instability in the electrical performance. Therefore, we recommend that thermodynamics be considered (e.g. ternary metal-Si-C phase diagrams) when selecting metallization schemes for high-temperature electronics. In addition to the thermodynamics, other requirements substantially complicate the task of developing high-temperature metallization schemes. For example, unless the device is hermetically sealed, the contacts must be highly resistant to oxidation. Other requirements include sufficient electrical conductivity, the ability to be connected to the outside world (e.g. wire bonding), and appropriate interfacial electrical properties (e.g. contact resistance or Schottky barrier height). In order to meet all of these requirements, multiple layers are typically combined. The additional layers introduce thermodynamic instability to what may have originally been a stable metal/SiC interface (e.g. TaC/SiC). In this case, one must consider diffusion barriers, which substantially slow the kinetics of diffusion/reaction between adjacent layers. Although there is no public record of a metal contact scheme that simultaneously meets all of these requirements, a number of single- and multi-layer contacts have been investigated. Some of the metal carbides and silicides appear promising.

Similar challenges also pertain to the thermal stability of Schottky contacts. We found that W-based (W, WC, WN, TiW) and Ni contacts were frequently investigated as high-temperature Schottky contacts, whereas Pd has been frequently investigated for high-temperature gas sensors because of its sensitivity to hydrogen. Tungsten-based Schottky contacts are potentially useful at higher temperatures than nickel contacts because of the higher reaction temperatures of the former, whereas Ni is advantageous because it yields a high Schottky barrier height on n-type SiC and can be used for both ohmic and Schottky contacts.

The formation of reproducible, thermally stable and low-resistance ohmic contacts to p-type SiC remains a critical problem for reliable performance of certain SiC-based devices. Because of its large bandgap and high work function, it is difficult to form ohmic contacts to p-type material by reducing the Schottky barrier height at the interface between the metal and the SiC. Reduction of the barrier width via high doping at the surface has been the preferred method for obtaining ohmic contacts to p-type SiC. Annealing the contacts at high temperatures (> 700°C) is also normally required.

The traditional approach to forming ohmic contacts to p-type SiC is to anneal an Al-based contact (e.g. TiAl) on p+ SiC at temperatures between 900 and 1150°C. However Al-based contacts are often considered to have problems with morphology and oxidation resulting from high processing and/or operating temperatures. It has been reported that a 70/30 wt% Al-Ti alloy showed the best results in terms of consistent contact resistances and reduced spiking and that ohmic behavior may be associated with melting.[69,70]

Because of the problems associated with Al, many researchers have investigated alternative p-type contacts. Perhaps the most successful alternatives are based on Pt. The advantages of Pt for p-type contacts include its high work function, its high melting point, and its oxidation resistance. Some groups have reported notable improvements via the incorporation of Si in the contacts to reduce or eliminate the formation of free C as a product of the contact anneal process. Single-phase (PtSi) contacts can be formed under controlled conditions and have the advantages that include an abrupt interface, the absence of free C, a reduced contact resistance and the improved stability at elevated temperatures.[14]

Nickel contacts, which are the most common ohmic contacts to n-type SiC, have also been shown to benefit from the incorporation of Si. Although annealed Ni on SiC produces ohmic contacts with a contact resistance as low as 1 x 10^{-5} ohm-cm^2, the reaction of Ni with SiC during the high temperature annealing causes undesirable morphological features such as interfacial roughening and carbon segregation. Presently the research focus is aimed at reducing or eliminating these morphological features in order to improve the long-term reliability and to better understand the mechanisms of ohmic contact formation. Alloying of the Ni layer with elements such as Al and Si has shown to produce an improved morphology, particularly when a reaction with the substrate is eliminated. These results have been achieved for both Ni_2Si and $NiSi_2$.[84,95,110,111]

The main technological challenge to the realization of large-area SiC power devices is the identification and reduction of electrically-active defects in the SiC crystals. In the early to mid-1990s, micropipes were considered the primary culprit that limited device performance. Micropipes were shown to consistently cause premature breakdown in 4H- and 6H-SiC p-n diodes.[112] Now that commercial suppliers have been able to substantially reduce the micropipe densities, electrical degradation due to other crystal defects have come under consideration. There is evidence that certain defects, such as closed-core screw dislocations and stacking faults, have deleterious effects on Schottky diodes. For example, the number of screw dislocations was shown to be inversely correlated with the breakdown voltage in 4H-SiC Schottky diodes.[113] Another study[116] indicates that the ideality factors and barrier heights may be affected by other unidentified defects. The results suggest that further studies are needed to elucidate the true nature of defects that affect the electrical properties of SiC Schottky diodes.

Acknowledgments

The authors would like to thank Mikael Ostling, Masataka Satoh, Brian Skromme, and Quamar Wahab for sending figures that were included in this chapter. Thanks are also expressed to the authors who sent us publications. Support from the National Science Foundation via Career Award No. ECS-9875186 is also gratefully acknowledged.

REFERENCES

1. Porter, L.M. and Davis, R.F., *Mater. Sci. Eng.*, **B34**, (1995), pp.83-105.
2. Katulka, G.L., Kolodzey, J. and Olowolafe, J., *IEEE Trans. Magn.*, **35**, (1999), pp.356-360.
3. Savage, S. and Spetz, A.L., *Comp. Semicond.*, **6**, (2000), pp.76-79.
4. Park, J.S., Landry, K. and Perepezko, J.H., *Mater. Sci. Eng.*, **A259**, (1999), pp.279-286.
5. Goesmann, F. and Schmid-Fetzer, R., *Mater. Sci. Eng.*, **B46**, (1997), pp.357-362.
6. Seng, W.F. and Barnes, P.A., *Mater. Sci. Eng.*, **B72**, (2000), pp.13-18.
7. Delucca, J.M. and Mohney, S.E., in *III-N Nitride, SiC and Diamond Materials for Electronic Devices,* Gaskill, D.K., Brandt, C., Nemanich, R.J., Eds. (Materials Research Society, 1996), vol. 423, pp.137-142.
8. Jang, T., Rutsch, G.W.M., Odekirk, B. and Porter, L.M., *Appl. Phys. Lett.*, **75**, (1999), pp.3956-3958.
9. Schuster, J.C., *Int. J. of Refractory Metals & Hard Materials,* **12**, (1993-1994) pp.173-177.
10. Kriz, J., Scholz, T., Gottfried, K., Leibelt, J., Kaufmann, C. and Gessner, T., *Mater. Sci. Forum,* **264-268**, (1998), pp.775-778.
11. Lee, S.-K., Zetterling, C.-M., Östling, M., Palmquist, J.-P., Högberg, H. and Jansson, U., *Solid-State Electron.*, **44**, (2000), pp.1179-1186.
12. Lee, S.-K., Zetterling, C.-M., Östling, M., Palmquist, J.-P. and Jansson, U., *Microelectronic Eng.*, **60**, (2002), pp.261-268.
13. Wenzel, R., Goesmann, F. and Schmid-Fetzer, R., *J. Mater. Sci: Materials in Electronics,* **9**, (1998), pp.109-113.
14. Jang, T., Erickson, J.W. and Porter, L.M., *J. Electron. Mater.*, **31**, (2002), pp.506-511.
15. Massalski, T.B., Okamoto, H., Subramanian, P.R. and Kacprzak, L., Eds., *Binary Alloy Phase Diagrams*, vol. 1 (ASM International, Materials Park, Ohio, 2nd ed., 1990).
16. Goesmann, F., Wenzel, R. and Schmid-Fetzer, R., *J. Mater. Sci.: Materials in Electronics,* **9**, (1998), pp.103-107.
17. Jang, T., Odekirk, B., Madsen, L.D. and Porter, L.M., *J. Appl. Phys.*, **90**, (2001), pp.4555-4559.

18. Okojie, R.S., Lukco, D., Chen, Y.L. and Spry, D.J., *J. Appl. Phys.*, **91**, (2002), pp.6553-6559.
19. Lundberg, N. and Ostling, M., *Solid-State Electron.*, **39**, (1996), pp.1559-1565.
20. Hwang, J.D., Fang, Y.K. and Song, Y.J., *Thin Solid Films,* **283**, (1996), pp.8–11.
21. Liu, S., Reinhardt, K., Severt, C. and Scofield, J., in *Silicon Carbide and Related Materials 1995. Proceedings of the Sixth International Conference;* Nakashima, S., Matsunami, H., Yoshida, S., Harima, H., Eds. (IOP Publishing Ltd, 1995), pp.589-592.
22. Constantinidis, G., Kornilios, N., Zekentes, K., Stoemenos, J. and Cioccio, L.D., *Mater. Sci. Eng.,* **B46**, (1997), pp.176-179.
23. Luckowski, E.D., Delucca, J.M., Williams, J.R., Mohney, S.E., Bozack, M.J., Isaacs-Smith, T. and Crofton, J., *J. Electron. Mater.*, **27**, (1998), pp.330-334.
24. Oder, T.N., Williams, J.R., Bozack, M.J., Iyer, V., Mohney, S.E. and Crofton, J., *J. Electron. Mater.*, **27**, (1998), pp.324-329.
25. Liu, S. and Scofield, J., in *Trans. of the Fourth Int. High Temp. Electronics Conf.* (IEEE, 1998), pp.88-92.
26. Crofton, J. et al., in *Trans. of the Fourth Int. High Temp. Electronics Conf.* (IEEE, 1998), pp.84-87.
27. Liu, S., *Mater. Sci. Forum* **264-268**, (1998), pp.791-794.
28. Gottfried, K., Fritsche, H., Driz, J., Leibelt, J., Kaurmann, C., Rudolf, F. and Gessner, T., *Mater. Sci. Forum*, **264-268**, (1998), pp.795-798.
29. Okojie, R.S., Ned, A.A., Kurtz, A.D. and Carr, W.N., *IEEE Trans. Electron. Dev.*, **46**, (1999), pp.269-274.
30. Jang, T., *Investigation of Thermally Stable Ohmic Contact Formation and Degradation Mechanisms on N-Type and P-Type SiC*. Ph.D. Thesis, Carnegie Mellon University (2002).
31. Touati, F., Kiminori, T. and Saji, M., *J. Electron. Mater.*, **28**, (1999), pp186-189.
32. Kassamakova, L. *et al.*, *IEEE Trans. Electron. Dev.*, **46**, (1999), pp.605-611.
33. Schmid, U., Getto, R., Sheppard, S.T. and Wondrak, W., *J. Appl. Phys.*, **85**, (1999), pp.2681-2686.
34. Kassamakova, L., Kakanakov, R., Kassamakov, I., Nordell, N., Savage, S., Svedberg, E.B. and Madsen, L.D., *Mater. Sci. Forum,* **338-342**, (2000), pp.1009-1012.
35. Lee, S.-K., Zetterling, C.-M., Danielsson, E. and Östling, M., *Appl. Phys. Lett.*, **77**, (2000), pp.1478-1480.
36. Lee, S.K., Zetterling, C.-M. and Östling, M., in *Silicon Carbide – Materials, Processing and Devices,* Agarwal, A., Skowronski, M., Cooper, J., Janzen, E., Eds. (Materials Research Society, Warrendale, PA, 2001), vol. 640, H7.2.1-H7.2.6.
37. Luckowski, E.D., Williams, J.R., Bozack, M.J., Isaacs-Smith, T. and Crofton, J., *Mat. Res. Soc. Symp. Proc.*, **423**, (1996), pp.119-124.
38. Saxena, V., Su, J.N. and Steckl, A.J., *IEEE Trans. Electron. Dev.*, **46**, (1999), pp.456-464.
39. Su, J.N. and Steckl, A.J., *Inst. Phys. Conf. Ser.*, **142**, (1996), pp.697-700.
40. Chen, L.-Y., Hunter, G.W., Neudeck, P.G. and Knight, D., *J. Vac. Sci. Technol.*, **A16**, (1998), pp.2890-2895.

41. Kim, C.K., Lee, J.H., Lee, Y.-H., Cho, N.I., Kim, D.J. and Kang, W.P., *J. Electron. Mater.*, **28**, (1999), pp.202-205.
42. Lundberg, N., Tågström, P., Jansson, U. and Östling, M., *Inst. Phys, Conf. Ser.*, **142**, (1996), pp.677-680.
43. Lundberg, N., Östling, M., Tägström, P. and Jansson, U., *J. Electrochem. Soc.*, **143**, (1996), pp.1662-1667.
44. Kassamakova, L. *et al.*, *Semicond. Sci. Technol.*, **13**, (1998), pp.1025-1030.
45. Kakanakova-Georgieva, A. et al., *Appl. Surf. Sci.*, **151**, (1999), pp.225-232.
46. Lundberg, N., Östling, M., Zetterling, C.-M., Tågström, P. and Jansson, U., *J. Electron. Mater.*, **29**, (2000), pp.372-375.
47. Shalish, I., Oliveira, C.E., Shapira, Y., Burstein, L. and Eizenberg, M., *J. Appl. Phys.*, **88**, (2000), pp.5724-5728.
48. Lee, S.-K., Zetterling, C.-M. and Östling, M., *J. Appl. Phys.*, **87**, (2000), pp.8039-8044.
49. Lee, S.-K., Zetterling, C.-M. and Östling, M., *J. Electron. Mater.*, **30**, (2001), pp.242-246.
50. Zhang, Q. and Sudarshan, T.S., *J. Electron. Mater.*, **30**, (2001), pp.1466-1470.
51. Raynaud, C., Isoird, K., Lazar, M., Johnson, C.M. and Wright, N., *J. Appl. Phys.*, **91**, (2002), pp.9841-9847.
52. Shalish, I., Gasser, S., Kolawa, E. and Nicolet, M.-A., in *Trans. of the Third Int. High Temp. Electronics Conf.* (IEEE, 1996), vol. 1, VII21-VII26.
53. Fomenko, V.S., *Handbook of Thermionic Properties*. Samsonov, G.V., Ed. (Plenum Press Data Division, New York, 1966).
54. Defives, D., Noblanc, O., Dua, C., Brylinski, C., Barthula, M. and Meyer, F., *Mater. Sci. Eng.*, **B61-62**, (1999), pp.395-401.
55. Skromme, B.J., Luckowski, E., Moore, K., Bhatnagar, M., Weitzel, C.E., Gehoski, T. and Ganser, D., *J. Electron. Mater.*, **29**, (2000), pp.376-383.
56. Waldrop, J.R. and Grant, R.W., *Appl. Phys. Lett.*, **56**, (1990), pp.557-559.
57. Waldrop, J.R., Grant, R.W., Wang, Y.C. and Davis, R.F., *J. Appl. Phys.*, **72**, (1992), pp.4757-4760.
58. Porter, L.M., Bow, J.S., Glass, R.C., Kim, M.J., Carpenter, R.W. and Davis, R.F., *J. Mater. Res.*, **10**, (1995), pp.668-679.
59. Itoh, A. and Matsunami, H., *Phys. Stat. Solidi, A* **162**, (1997), pp.389.
60. Pelletier, J., Gervais, D. and Pomot, C., *J. Appl. Phys.*, **55**, (1984), pp994-1002.
61. Spiel, L., Nennewithz, O. and Pezoldt, J., *Inst. Phys. Conf. Ser.*, **142**, (1996), pp.585-588.
62. Nordell, N., Savage, S. and Schoner, A., *Inst. Phys. Conf. Ser.*, **142**, (1996), pp.573-576.
63. Iliadis, A.A. *et al.*, *Appl. Phys. Lett.*, **73**, (1998), pp.3545-3547.
64. Zhao, J.H., Tone, K., Weoner, S.R., Caleca, M.A., Du, H. and Withrow, S.P., *IEEE Electron. Dev. Lett.*, **18**, (1997), pp.375-377.
65. Crofton, J., Beyer, L., Williams, J.R., Luckowski, E.D., Mohney, S.E. and Delucca, J.M., *Solid-State Electron.*, **41**, (1997), pp.1725-1729.
66. Papanicolaou, N.A., Edwards, A., Rao, M.V. and Anderson, W.T., *Appl. Phys. Lett.*, **73**, (1998), pp.2009-2011.

67. Jang, T., Rutsch, G., Odekirk, B. and Porter, L.M., *Mater. Sci. Forum,* **338-342**, (2000), pp.1001-1004.
68. Oder, T.N., Williams, J.R. and Mohney, S.E., *J. Electron. Mater.,* **27**, (1998), pp.12-16.
69. Crofton, J., Mohney, S.E., Williams, J.R. and Isaacs-Smith, T., *Solid-State Electron.,* **46**, (2002), pp.109-113.
70. Mohney, S.E., Hull, A., Lin, J.Y. and Crofton, J., *Solid-State Electron.,* **46**, (2002), pp.689-693.
71. Vassilevski, K.V., Constantinidis, G., Papanicolaou, N., Martin, N. and Zekentes, K., *Mater. Sci. Eng.,* **B61-2**, (1999), pp.296-300.
72. Kakanakov, R., Kassamakova, L., Kassamakov, I., Zekentes, K. and Kuznetsov, N., *Mater. Sci. Eng.,* **B80**, (2001), pp.374-377.
73. Nakatsuka, O., Takei, T. and Koide, Y., *Mater. Trans.,* **43**, (2002), pp.1684-1688.
74. Nakatsuka, O., Koide, Y. and Murakami, M., *Mater. Sci. Forum,* **389-393**, (2002), pp.885-888.
75. Kassamakova, L., Kakanakov, R. and Kassamakov, I., *Mater. Sci. Forum,* **338-342**, (2000), pp.1009-1012.
76. Kassamakova, L., Kakanakov, R. and Nordell, N., *Mater. Sci. Forum,* **264-268**, (1998), pp.787-790.
77. Kassamakova, L., Kakanakov, R., Nordell, N., Savage, S., Kakanakova-Georgieva, A. and Marinova, T., *Mater. Sci. Eng.,* **B61-62**, (1999), pp.291-295.
78. Crofton, J., Porter, L.M. and Williams, J.R., *Phys. Stat. Solidi,* **B202**, (1997), pp.581-603.
79. Jung, K.H., Cho, N.I. and Lee, J.H., *Mater. Sci. Forum,* **389-393**, (2002), pp.905-908.
80. Fursin, L.G., Zhao, J.H. and Weiner, M., *Electron. Lett.,* **37**, (2001), pp.1092-1093.
81. Glass, R.C., Palmour, J.W., Davis, R.F. and Porter, L.M., *"Method of forming ohmic contacts to p-type wide bandgap semiconductors and resulting ohmic contact structure"* U.S.A. Patent No. 5,323,022 (1994).
82. Crofton, J., McMullin, P.G., Williams, J.R. and Bozack, M.J., *J. Appl. Phys.,* **77**, (1995), pp.1317-1319.
83. Uemoto, T., *Jpn. J. Appl. Phys.,* **34**, (1995), pp.L7-L9.
84. Nakamura, T. and Satoh, M., *Solid-State Electron.,* **46**, (2002), pp.2063-2067.
85. Levit, M., Grimberg, I. and Weiss, B.Z., *J. Appl. Phys.,* **80**, (1996), pp.167-173.
86. Marinova, T., Yakimova, R. and Krastev, V., *J. Vac. Sci. Tech.,* **B14**, (1996), pp.3252-3256.
87. Marinova, T., Krastev, V. and Hallin, C., *Mater. Sci. Forum,* **207-209**, (1996), pp.293-296.
88. Rastegaeva, M.G., Andreev, A.N. and Petrov, A.A., *Mater. Sci. Eng.,* **B46**, (1997), pp.254-258.
89. Hallin, C., Yakimova, R. and Pecz, B., *J. Electron. Mater.,* **26**, (1997), pp.119-122.
90. Jacob, C., Pirouz, P. and Kou, H.I., *Solid-State Electron.,* **42**, (1998), pp.2329-2334.
91. Cole, M.W., Joshi, P.C. and Hubbard, C.W., *J. Appl. Phys.,* **88**, (2000), pp.2652-2657.

92. Oskam, G., Searson, P.C. and Cole, M.W., *Appl. Phys. Lett.,* **76**, (2000), pp.1300-1302.
93. Han, S.Y. *et al., Appl. Phys. Lett.,* **79**, (2000), pp.1816-1818.
94. Cole, M.W., Joshi, P.C. and Ervin, M., *J. Appl. Phys.,* **89**, (2001), pp.4413-4416.
95. Nakamura, T., Shimada, H. and Satoh, M., *Mater. Sci. Forum,* **338-342**, (2000), pp.985-8.
96. Cole, M.W., Joshi, P.C. and Hubbard, C., *J. Appl. Phys.,* **91**, (2002), pp.3864-3868.
97. Kurimoto, E., Harima, H. and Toda, T., *J. Appl. Phys.,* **91**, (2002), pp.10215-10217.
98. Han, S.Y., Shin, J.Y. and Lee, B.T., *J. Vac. Sci. Tech.,* **B20**, (2002), pp.1496-1500.
99. Via, F.L., Roccaforte, F. and Makhtari, A., *Microelectronic Eng.,* **60**, (2002), pp.269-282.
100. Toda, T., Ueda, Y. and Sawada, M., *Mater. Sci. Forum,* **338-342**, (2000), pp.989-992.
101. Madsen, L.D., Svedberg, E.B., Radamson, H.H., Hallin, C. and Jorvarsson, B.H., *Mater. Sci. Forum,* **338-42**, (2000), pp.981-984.
102. Basin, Y.M., Kuznetsov, V.N., Markov, V.T. and Guzei, L.S., *Russ. Met.,* **4**, (1988), pp.197-200.
103. Goesmann, F. and Schmid-Fetzer, S., *Mater. Sci. Eng.,* **B46**, (1997), pp.357-362.
104. Robbie, K., Jemander, S.T. and Lin, N., *Mater. Sci. Forum,* **339-342**, (2000), pp.981-984.
105. Han, S.Y., Kim, N.K. and Kim, E.D., *Mater. Sci. Forum,* **389-93**, (2002), pp.897-900.
106. Bachli, A., *Mater. Sci. Eng.,* **B56**, (1997), pp11-23.
107. Roccaforte, F., Via, F.L. and Raineri, V., *Mater. Sci. Forum,* **389-393**, (2002), pp.893-896.
108. Tanimoto, S., Kiritani, N. and Hoshi, M., *Mater. Sci. Forum,* **389-393**, (2002), pp.879-884.
109. Lu, W., Mitchel, W.C., Thornton, C.A., Landis, G.R. and Collins, W.E., *J. Electron. Mater.,* **32**, (2003), pp.426-421.
110. Nakamura, T., Shimada, H. and Satoh, M., *Mater. Sci. Forum,* **338-342**, (2000), pp.985-988.
111. Nakamura, T. and Satoh, M., *Mater. Sci. Forum,* **389-93**, (2002), pp.889-892.
112. Neudeck, P.G. and Powell, J.A., *IEEE Electron Dev. Lett.,* **15**, (1994), pp.63-65.
113. Wahab, Q., Ellison, A., Henry, A., Janzen, E., Hallin, C., Persio, J.D. and Martinez, R., *Appl. Phys. Lett.,* **76**, (2000), pp.2725-2727.
114. Zimmermann, U., Österman, J., Kuylenstierna, D., Hallen, A., Konstantinov, A.O., Vetter, W.M. and Dudley, M., *J. Appl. Phys.,* **93**, (2003), pp.611-618.
115. Schnabel, C.M., Tabib-Azar, M., Neudeck, P.G., Bailey, S.G., Su, H.B., Dudley, M. and Raffaelle, R.P., *Mater. Sci. Forum,* **338-342**, (2000), pp.489-492.
116. Skromme, B.J. *et al.*, in *Silicon Carbide 2002 – Materials, Processing and Devices* Saddow, S.E., Larkin, D.J., Saks, N.S., Schoener, A., Skowronski, M., Eds. (Materials Research Society, Warrendale, PA, 2003), vol. 742, K3.4.1-K3.4.6.
117. Tung, R.T., *Appl. Phys. Lett.,* **58**, (1991), pp.2821-2823.
118. Tung, R.T., *Phys. Rev.,* **B45**, (1992), pp.13509-13523.
119. Tung, R.T., *J. Vac. Sci. Technol.,* **B11**, (1993), pp.1546-1552.

120. Skromme, B.J., Luckowski, E., Moore, K., Clemens, S., Resnick, D., Gehoski, T. and Ganser, D., *Mater. Sci. Forum,* **338-342**, (2000), pp.1029-1032.
121. Deeb, C., Kahn, H., Milhet, X., Zorman, C., Mehregany, M. and Heuer, A.H., in *Silicon Carbide – Materials, Processing and Devices,* Agarwal, A., Skowronski, M., Cooper, J., Janzen, E., Eds. (Materials Research Society, Warrendale, PA, 2001), vol. 640, H5.22.1-H5.22.5.
122. Liu, J.Q., Skowronski, M., Hallin, C., Söderholm, R. and Lendenmann, H., *Appl. Phys. Lett.,* **80**, (2002), pp.749-751.
123. Bergman, J.P., Lendenmann, H., Nilsson, P.Å., Lindefelt, U. and Skytt, P., *Mater. Sci. Forum,* **353-356**, (2001), pp.299-302.
124. Lendenmann, H., Dahlquist, F., Johansson, N., Söderholm, R., Nilsson, P.Å., Bergman, J.P. and Skytt, P., *Mater. Sci. Forum,* **353-356**, (2001), pp.727-730.
125. Okojie, R.S., Xhang, M., Pirouz, P., Sergey, T., Jessen, G. and Brillson, L.J., *Appl. Phys. Lett.,* **79**, (2001), pp.3056-3058.
126. Skromme, B.J. *et al.*, *Mater. Sci. Forum,* **389-393**, (2002), pp.455-458.

CHAPTER 4

DRY ETCHING OF SIC

S.J. Pearton

Department of Material Science and Engineering,
University of Florida
Gainesville, FL 32611 USA
spear@mse.ufl.edu

In this chapter we discuss wet and dry patterning techniques for SiC and the relative merits of these methods for MEMS processing. We describe the basic principles involved in etching SiC and problems that can arise because of the binary nature of the lattice and its relatively high bond strength. Recent developments in the use of high density plasma sources to achieve fast etching rates (in some cases over 1 $\mu m\ min^{-1}$ for bulk 4H-SiC) are discussed. These sources are likely to play a dominant role for processing of SiC MEMS devices since they are capable of producing etch depths from 0.1 to 100 μm with minimal disruption of the SiC surface.

1. Introduction

Due to its hardness ($H=9^{+}$), SiC is one of the most widely used lapping and polishing abrasives for metals, metallic components and semiconductor wafers. However this very property makes it difficult to etch in typical acid or base solutions. In its single crystal form, SiC is not attacked by single acids at room temperature. Indeed the only techniques for etching SiC employ molten salt fluxes, hot gases, electrochemical processes or plasma etching.[1-11] Table 1 shows a list of the molten salt solutions and the temperatures needed for successful etching of SiC. The disadvantages of these high temperature, corrosive mixtures include the need for expensive

Pt beakers and sample holders (which can withstand the molten salt solutions) and the inability to etch masked samples because few masks hold up to these mixtures. While one can conceivably use Pt masks, the wet etching is isotropic and therefore undercuts the mask.

Solution	**Material**	**Temperature**	**Reference**
NaF/K_2CO_3	SiC(0001)	650°C	[3]
H_3PO_4	a-SiC(H)	180°C	[4]
NaOH	SiC(111)	900°C	[5]
Na_2O_2	SiC(0001)	>400°C	[6]
$NaOH/Na_2O_2$	SiC(0001)	700°C	[7]
Borax/Na_2CO_3	epi-SiC	855°C	[8]
NaOH/KOH	bulk 6H	480°C	[9]
$Na_2O_2/NaNO_2$	bulk 6H	>400°C	[10]
KOH/KNO_3	bulk 6H	350°C	[11]

Table 1. Molten flux and other wet etchants for SiC

Photoelectrochemical etching can be successfully employed for SiC.[12] The dissolution rate of semiconductors may be altered in acid or base solutions by illumination with above bandgap light. The mechanism for photo-enhanced etching involves the creation of e-h pairs, the subsequent oxidative dissociation of the semiconductor into its component elements (a reaction that consumes the photo-generated holes) and the reduction of the oxidizing agent in the solution by reaction with the photo-generated electrons. Generally, n-type material is readily etched under these conditions, while p-type material is not due to the requirements for confining photo-generated holes at the semiconductor-electrolyte interface (i.e. the p-surface is depleted of holes because of the band-bending). This allows for selective removal of n-SiC from an underlying p-SiC layer.[12] Under conditions of no illumination, it is often possible to get the reverse selectivity if the sample is correctly biased, since n-SiC requires photogeneration of carriers for etching to proceed. Etching over large areas can be achieved using Hg lamps and some degree of anisotropy is obtained because of the shadowing effect of the metal masks (typically Ti) allowing carriers to be generated only in unmasked regions. Some of the disadvantages of the technique include fairly rough surface morphologies (due to enhanced dissolution rates for areas around crystal defects), inability to pattern very small dimension features and poor uniformity of etch rate. For these reasons, most attention is now focussed on dry etching methods for SiC, most of which have been developed for high power, high temperature electronics in this materials system.[14-53]

2. Basics of Plasma Etching

A plasma is an ionized gas with equal numbers of free positive and negative charges. The free charge is produced by the passage of electric current through the discharge. For most plasma of interest for etching, the extent of ionization is very small. Typically there is only one charged particle per 100,000 to 1,000,000 neutral atoms and molecules. The positive charge is mostly in the form of singly ionized neutrals, (i.e. atoms, radicals or molecules) from which a single electron has been stripped (removed). The majority of negatively charged particles are usually free electrons; although in very electronegative gases such as chlorine, negative ions can be more abundant.

2.1. Plasma creation

For the generation of plasma, a high frequency voltage is applied between the two electrodes. Free electrons (current flows) are accelerated and collide with neutral gas. When the collision occur, the energy of the gas molecules becomes high and the molecules can then be dissociated, ionized and excited.

Dissociation : $e + XY \longrightarrow X + Y + e$
Ionization : $e + XY \longrightarrow XY^+ + 2e$
Excitation : $e + XY^+ \longrightarrow X + Y^*$

* indicates an excited atom or radical.

In a plasma, the electron and ion densities are equal on average, but less than the density of neutral species. When a plasma is created, electrons and ions will diffuse out of the plasma. The electrons will reach the surface of the material which is exposed in the plasma before the ions do due to their (electrons) much greater velocity. This causes the plasma to become more positive, since there is an excess of positive ions left behind. The surfaces of the plasma containment vessel charge up negatively. This negative charge pushes other electrons away at the same time as attracting positive ions. In steady state the surface no longer charges up, and thus electrons and positive ions have to arrive at the same rate. The field near the surface holds the electrons away from the surface, allowing only the most energetic electrons to get there. The field also accelerates the positive ions toward the surface, and in this way the rates of arrival of electrons and positive ions are made equal.

In particular on the powered electrode, electrons are excluded by the positive charge, producing a region above the electrode in which there are fewer collisions of gas molecules with energetic electrons. For this reason, this region appears dark compared to the rest of the plasma region due to the absence of emission from excited molecules. This positively charged region is called a *sheath*.

2.2. Basic mechanism of plasma etching

In plasma, two kinds of active species are produced, neutrals and ions. Neutrals may be very reactive,while the ions are usually less reactive but their kinetic energy can be controlled by substrate bias.

2.2.1. Sputtering

Positive ions are accelerated through sheath region and strike the substrate with high kinetic energy[54]

$$E_{max} = \frac{(QE_0)^2}{8\pi^2 f^2 M} eV$$

where Q is the ionic charge, E_0 represents the rf field (kV/cm), M is the mass of the ion and f is the rf frequency (MHz).

By momentum conservation law, some of this energy is transferred to surface atoms that are then ejected, leading to material removal. This is mechanical interaction and the sputtering rate is given by[55]

$$R = \frac{62.2 sjW}{\rho}$$

where s is etch yield, j is ion flux (mA/cm^2), W is atomic weight (g/mol) and ρ is material density (g/cm^3).

Sputtering is unselective etching because the ion energy required to eject material is large compared to differences in surface bond energies and chemical reactivity. Due to the applied voltage to the substrate, the flux of ions is vertical and this kind of etching is anisotropic.

2.2.2. Chemical reaction

This etching comes about when active species (neutral) from the gas phase are absorbed on the surface material and react with it to form a volatile product. High product volatility is essential. The evaporation rate of a material is given by[55]

$$\mu_A = \alpha \left[\frac{M}{2\pi RT} \right]^{-\frac{1}{2}} P$$

where α is the material-dependent efficiency factor, usually between 0.1 and 1.0, M is the molecular weight and P is the vapor pressure.

Without volatility the reaction products would coat the surface and prevent gaseous species from reacting it, and cut off the etching reaction. Chemical etching provides very high selectivity but is non-directional, producing isotropic etching.

2.2.3. Ion-assisted plasma etching

The substrates are exposed to suitable neutral species in the presence of ion-bombardment. The combination between sputtering and chemical reaction results in material removal rates exceeding the sum of separate chemical attack and sputtering. There are two mechanisms for ion-assisted etching.

(1) Ion-enhanced energetic etching

This etching, neutral species cause little or no etching without ion bombardment. Ions damage the substrate material create high roughness which increased the expose surface, dangling the atomic bond of the surface which increase the number of absorption site. Since ions are accelerated and strike surface vertically, the etching induced is directional.

(2) Ion-enhanced inhibitor

Etching by neutrals is spontaneous so ion bombardment does not cause the etching reaction. Ion can coat substrate surface and prevent etching reaction from taking place. The normal going ion flux keeps areas clear of film on the horizontal surfaces, while vertical feature sidewalls are coated with a thin film which inhibits chemical reaction.

(3) Plasma etching parameters

(a) Effect of pressure

Pressure is inversely proportional to the mean free path of particle. At higher pressure, the mean free path is shorter cause more frequency of collision of electrons. The electrons will lose their energy and create more reactive neutral species during the collision (generate higher plasma density). Then the etch mechanism is dominated by chemical reaction rather than physical (sputtering) reaction.

As pressure is lowered, the characteristic potentials across the sheaths and the voltage applied to a discharge increase sharply. The rise in potential translates into a higher energy ion flux to substrate surfaces. Sputtering does not take place until ion energy exceeds the material-ion (molecule) threshold energy.

(b) Effect of temperature

Temperature is a function of chemical reaction as $e^{-Ea/RT}$, where E_a is an activation energy, R is gas constant and T is temperature in Kelvin. Thus, it has a dominant effect on selectivity, etch rates and the degradation on resist mask.

(c) Effect of loading

The loading effect is the decreasing of etch rate when there are more etchable substrate material placed in a reactor. The etch rate is usually proportional to etchant concentration, their concentration decreases with the area of etchable surface in the plasma.

2.3. Plasma reactors

2.3.1. Ion milling

Ion milling is a pure physical process. The commonly gas used is Ar. The ion energy and ion density is separated control by the filament current and the accelerated voltage adjustment. This process employs high energetic inert ion to erode the surface of material by bombardment, causing high surface damage and degrade the performance of the device.

2.3.2 Reactive Ion Etching (RIE)

Reactive Ion Etching technique generates the plasma at a radio frequency of 13.56 MHz between two parallel electrodes in a reactive gas (see Figure 1). The electrons will be accelerated and collide with gas molecules contribute to sustaining the plasma.

The substrate is placed on the power electrode, not grounded, in this case a large negative dc self-bias develop on the sample and attract ion from plasma which cause damage on the surface. This results to high etch rate and anisotropic. However, highly energetic ions damage the sample surface and degrade both electrical and optical device performances.

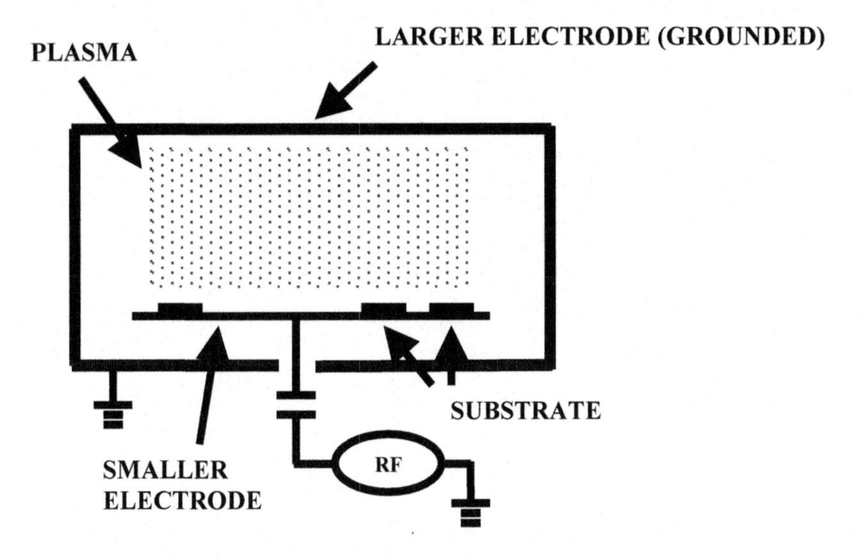

Figure 1. Schematic of RIE reactor.

2.3.3. Electron-Cyclotron Resonance (ECR)

Due to the surface damage from high energetic ions, High-density plasma is interesting. High-density ECR plasmas are formed at low pressures with low plasma potentials and ion energies due to magnetic confinement of electrons in the source region (Figure 2). Therefore, the surface damage in ECR may less than with the RIE technique if ion energy is the most important parameter in determining damage. In other cases, the higher ion flux may induce more damage than with RIE.

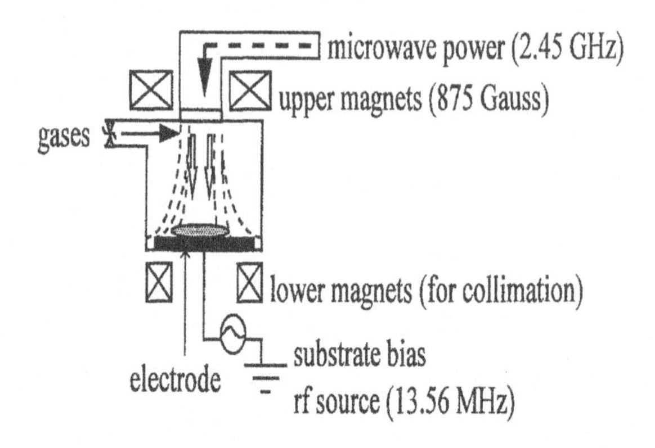

Figure 2. Schematic of ECR reactor.

The frequency of orbital motion of electrons, confined under the action of an external magnetic field (875 Gauss) is equivalent to the drive frequency of 2.45 GHz leading to the occurrence of resonance, called electron cyclotron resonance, if this frequency of power is applied to the plasma. In this condition, outer shell electrons from gas molecules in the discharge may also be liberated, leading to a very high degree of ionization in the plasma. ECR provides high ion density (10^{10}-10^{12} cm^{-3}) compared with RIE (10^9 cm^{-3}) without inducing high damage on the sample because the plasma potential is much lower.

ECR can also control the ion energy and ion flux independently. Ion energy is controlled by rf or dc biasing at the substrate holder while the ion and neutral flux is controlled by microwave and gas pressure.

2.3.4. Inductive Coupled-Plasma (ICP)

Inductive coupled-plasma etching offers an alternative high-density plasma technique where plasmas are formed in a dielectric vessel encircled by an inductive coil into which rf power is applied (see Figure 3). A strong magnetic field is induced in the center of the chamber, which generates a high-density plasma ($\sim 5 \times 10^{11}$ cm^{-3}) due to the circular region of the electric field that exists concentric to the coil. The electrons in circular path will have only a small chance to be lost to the chamber walls resulting in low dc self bias. At low pressures (≤ 20 mTorr), the plasma diffuses from the generation region and drifts to the substrate at relatively low ion energy (< 25 eV). Thus ICP etching is expected to produce low damage while achieving high etch rates. Anisotropic profiles are obtained by superimposing a rf bias on the sample to independently control ion energy and by using glow pressure conditions to minimize ion scattering and lateral etching. ICP sources may be easier to scale up than ECR sources and are more economical in terms of cost and power requirements.

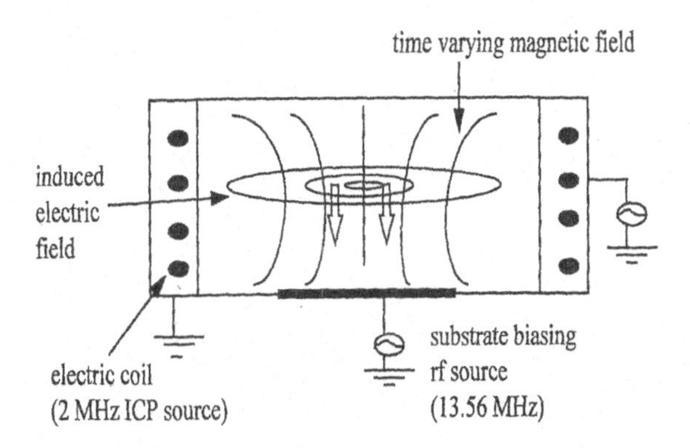

Figure 3. Schematic of ICP reactor.

3. Plasma Etching of Silicon Carbide

In order to etch silicon carbide in a plasma reactor, the chemistry used must be reactive with SiC and the species produced by the chemical reactions must be volatile compounds under the temperature and pressure condition to avoid the residue on the surface.

Several chemistries were examined (see Table 2). The most effective gas is based with fluorine chemistry. The reaction mechanism of SiC in F_2-based chemistry is shown below.

$$Si + xF \longrightarrow SiF_x \qquad x \leq 1\text{-}4$$
$$C + xF \longrightarrow CF_x \qquad x \leq 1\text{-}2$$

J.J. Wang *et al.*[52] showed from optical emission spectra that ion bombardment plays a role in the etch mechanism.

Reactor	SiC	Gas	Condition at highest ER	ER ($Å \cdot min^{-1}$)	Ref
RIE	6H	CHF_3/O	20sccm, 200W, 0% O_2	32	[13]
		SF_6/O_2		410	
		CF_4/O_2		278	
		NF_3/O_2		493	
	3C	SF_6	150W, 80mTorr	700	[14]
	6H,4H	NF_3	225mTorr, 95-110sccm, 275W	1500	[15]
ECR	6H	CF_4/O_2	5W, -100V, 17.5% O_2, 50sccm	800	[16]
	3C,6H	CF_4/O_2	650W, -100V, 17% O_2, 50sccm	700	[17]
	6H	$20SF_6/10Ar$	750W, 250rf, 2mTorr, 30sccm	4500	[18]
	3C,5H	SF_6/O_2	1200W, 1mTorr, 4sccm	2500	[19]
	6H	$10Cl_2/5Ar$	1000W,150rf,1.5mTorr,15sccm	2500	[20]
		$10Cl_2/5H_2$	1000W,150rf,1.5mTorr,15sccm	1000	
		4IBr/4Ar	1000W,250rf,1.5mTorr,15sccm	1100	
		NF_3	800W,100 rf,1 mTorr,10 sccm	1600	
		SF_6	800W,100 rf,1mTorr,10 sccm	450	
ICP	6H	Cl_2Ar or /He	100 Cl_2, 750W, 250 rf, 5 mTorr	100	[18]
		Cl_2/Xe	13% Cl_2,750W, 250 rf, 5 mTorr	260	
		IBr/Ar	10% Ar, 750W, 250 rf, 5 mTorr	800	
		ICl/Ar	66% Ar, 750W, 250 rf, 5 mTorr	250	
	6H	NF_3/O_2 or Ar	100% NF_3,750W,250rf,5 mTorr	4000	[21]
	4H	NH_3	100% NF_3, 500W, 50rf 2 mTorr	8000	[22]
	6H	SF_6	100% SF_6,900W,-450V,5mTorr	9700	[23]
Helicon	4H	$30SF_6/7/5\ O_2$	25% O_2, 200W, -500V,6 mTorr	13.500	[24]

Table 2. Published plasma etch rates of SiC.

Various gas addition can have effects on the etch behavior. Oxygen has often been added to fluorine-based chemistries under RIE conditions to enhance the active fluorine concentration and increase SiC etch rate. In ECR conditions, there is only little change in the atomic fluorine concentration.[48,49] In contrast, the addition of H_2 to the gas mixture reduces the etch rate.[13] The introduction of hydrogen into the plasma prevents residue formation through a combination of mechanism, including the formation of volatile alane (AlH_3) and the removal of the C-rich surface.[13]

The differences in the etch rates are due more to differences in the dangling bond densities and the corresponding reactivities of the crystal faces than to the different crystal structure. For example, each atom on cubic (001) face has two dangling bonds, whereas only one dangling bond exists on a (111) face or similarly to the (0001) face of hexagonal SiC.

There is no measurable difference in etch rates between n+ and p+ SiC indicating that Fermi level effects play no role in the etch mechanism under ICP conditions.[52] By contrast with the RIE technique, the etch rate increases when the n-type doping increases.[24]

In order to etch silicon carbide in a plasma reactor, the chemistry used must be reactive with SiC and the species produced by the chemical reactions must be volatile compounds under the operating temperature and pressure conditions to avoid residues on the surface.

Many plasma chemistries have been examined (see Table 2). The most effective gases in terms of etch rate are based on fluorine chemistry. The reaction mechanism of SiC in F_2-based chemistry is shown as below.

$$Si + xF \rightarrow SiF_x \qquad x \leq 1\text{-}4$$
$$C + xF \rightarrow CF_x \qquad x \leq 1\text{-}2$$

From optical emission spectra it is clear that ion bombardment play a role in the etch mechanism. When etching silicon atoms with atomic fluorine, a carbon layer is present on the exposed surface and is removed by the ion bombardment.

Various gas additions can have effects on the etch behaviour. Oxygen has often been added to fluorine-based chemistries under RIE conditions to enhance the active fluorine concentration and increase SiC etch rate. In high ion density conditions this produces only a small change in the atomic fluorine concentration. In contrast, the addition of H_2 to the gas mixture reduces the etch rate. The introduction of hydrogen into the plasma prevents residue formation through a combination of mechanisms, including the formation of volatile alane (AlH_3) to remove Al sputtered from the reactor and the removal of the C-rich surface.

The differences in the etch rates are due more to differences in the dangling bond densities and the corresponding reactivities of the crystal faces than to the different crystal structures. For example, each atom on cubic (001) face has two dangling bonds, whereas only one dangling bond exists on a (111) face or similarly on the (0001) face of hexagonal SiC.

Previous results on reactive ion etching of SiC have generally employed F_2-based plasmas.Relatively rough surfaces are often observed under these conditions due to sputtering of the electrode material onto the SiC sample, leading to micromasking.

With the advent of high density plasma sources, including Electron Cyclotron Resonance (ECR), Inductively Coupled Plasma (ICP) and Helicon, much higher SiC etch rates have been reported. The key advantage of these sources is decoupling of ion energy and ion flux, so that relatively low ion energies can be employs. Schematics of RIE, ECR and ICP reactors are shown in Figures 1-3. This reduces the electrode sputtering problem and in addition the plasma chemistries for high density sources generally involve gases that do not contain CH_x because of the extensive polymer deposition that can occur within the source at high applied powers. The absence of these two sources of redeposition onto the SiC generally leads to good surface morphologies.

Inductively coupled-plasma etching offers an attractive high-density plasma technique where plasmas are formed in a dielectric vessel encircled by an inductive coil into which rf power is applied. A strong magnetic field is induced in the center of the chamber, which generates a high-density plasma ($\sim 5\times10^{11}$ cm^{-3}) due to the circular region of the electric field that exists concentric to the coil. The electrons in this circular path will have only a small chance to be lost to the chamber walls, resulting in low dc self bias. At low pressures ($\leq$ 20 mTorr), the plasma diffuses from the generation region and drifts to the substrate at relatively low ion energy ($<$ 25 eV). Thus ICP etching is expected to produce low damage while achieving high etch rates. Anisotropic profiles are obtained by superimposing a rf bias on the sample to independently control ion energy and by using flow pressure conditions to minimize ion scattering and lateral etching. ICP sources may be easier to scale up the ECR sources and are more economical in terms of cost and power requirements.

In choosing the optimum plasma chemistries for investigation, it is instructive to look at SiC etch rates reported previously in the literature (Table 2). There are two key points evident in this data. Firstly the high density reactors do indeed produce faster rates, and secondly F_2-based chemistries lead to higher rates than Cl_2, F_2 or Br_2. This is readily understood by examining the relative volatility of the SiC etch products in F_2- or Cl_2-based plasmas. Table 3 shows the boiling points for potential etch products in these plasmas (with the addition of O_2 in both cases, although it is reported that O_2 itself plays no direct role in SiC etching but rather can influence the etch rate through changing the atomic fluorine neutral density in the discharge). While it is understood that the high ion flux in ICP discharges can desorb the etch products before they can fully coordinated, the boiling points of the complete molecules do give some indication of relative volatility and hence the trend expected for etch rates in the different chemistries. From Table 3, it is clear that the fluorinated products are more volatile than their chlorinated counterparts. Finally, the etching should have a high selectivity over both the mask material and the front-side metal employed as the etch-stop. The thickness of the SiC substrate enables us to estimate that an etch rate of at least 4000 Å·min^{-1} is needed to keep the process time below ~2 hours, which is a rough guess for a practical process.

Etching Product	Boiling Points (°C)
$SiCl_4$	57.6
SiF_4	-86
CCl_4	76.8
CF_4	-128
CO_2	-78.5 subl
CO	-191.5

Table 3. Boiling points of potential etch products in plasma etching of SiC.[57-60]

4. Plasma Chemistries

Figure 4 shows the etch rates (top) and etch yields (bottom) for SiC in ICP discharges of NF_3, SF_6, BF_3 or PF_5 at fixed rf chuck power (250W) and pressure (2 mTorr), as a function of ICP source power. The yield tends to decrease as the source power is increased, even as etch rate increases with NF_3, SF_6 and PF_5. This suggests that ion flux is not the limiting factor under these conditions, but rather the supply of fluorine neutrals to the SiC surface limits the etch rate. The etch rates are significantly higher with NF_3 and SF_6, which is consistent with the lower bond strength of these molecules compared to PF_5 and BF_3. When comparing the relative advantages of NF_3 and SF_6, the much lower cost of the latter outweighs the faster rates obtained with the former, particularly for long etch times.In these experiments the rf chuck power was held constant at 250W, corresponding to dc self-biases of roughly –290V at 250W source power, to –200V at 1500W source power. Clearly, NF_3 and SF_6 produce the fastest rates, and this correlates to the relative dissociation of these gases in the ICP source. Optical emission spectroscopy showed very intense atomic fluorine lines in the range 700-900 nm for both NF_3 and SF_6, while the intensities of these lines were much lower for BF_3 and PF_5. The etch rates are also in good correlation with the average bond energies for the feedstock gases, i.e. BF_3 154 kCal/mol,[57] PF_5 126 kCal/mol,[58] SF_6 78.3 kCal/mol[59] and NF_3 66.4 kCal/mol.[60] The lower the bond energy, the more effective is the dissociation in the ICP source to form atomic fluorine neutrals which are the active etchant species. The etch products are probably SiF_x and CF_x species (x would not necessarily have to reach its fully coordinated value of 4 under ion-assisted conditions), although we did not have adequate sensitivity in our OES system to detect them during the etching process. In the case of BF_3 the SiC etch rate decreases slightly at high source powers, which might be related to the fall-off in ion energy under those conditions or to desorption of the reactant fluorine before it can form etch products with the SiC.

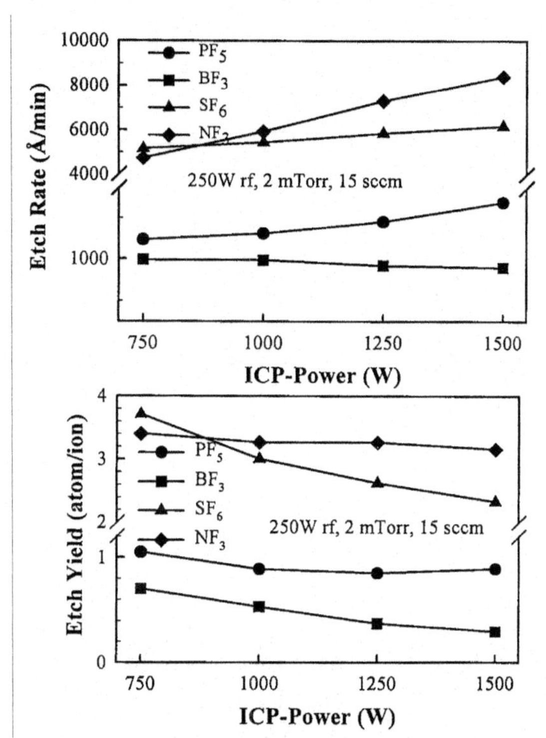

Figure 4. Etch rates (top) and etch yield (bottom) for SiC in different ICP discharges, as a function of source power.

Figure 5 shows the rf power dependence of SiC etch rate at a fixed ICP source power of 750W. The dc self-bias increases almost linearly with chuck power, as shown in the lower part of the figure. In NF_3, SF_6 and PF_5 there is a general trend for increasing etch rate as rf chuck power is increased. This could be related to increased Si-C bond-breaking efficiency at higher ion energies, allowing more etch products to form. In the case of BF_3 the etch-limiting step is probably the supply of atomic fluorine because of the lower efficiency in dissociating this gas.

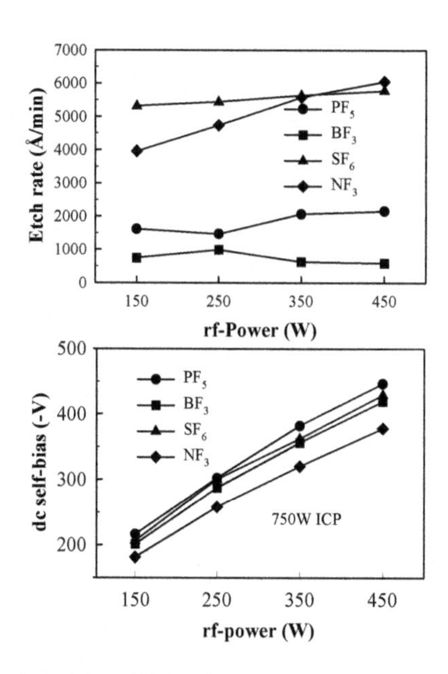

Figure 5. Etch rate (top) and dc self-bias (bottom) as a function of rf chuck power for SiC in pure NF_3, SF_6, PF_5 or BF_3 discharges (750W source power, 2 mTorr).

Polished SiC surfaces often have relatively rough morphologies due to residual mechanical damage. After dry etching with any of the different plasma chemistries, the surface roughness improves to values in the range 0.6-2.0 nm. This smoothing of initially rough surfaces is commonly observed in ion-driven etch processes and originates in the angular dependence of ion mill rates. This leads to faster removal rates for high aspect ratio features and creates a smoother morphology. Figure 6 shows the dependence of RMS surface roughness in ICP source power in the four different plasma chemistries. Under virtually all conditions the etched surfaces are smoother than the unetched control samples.

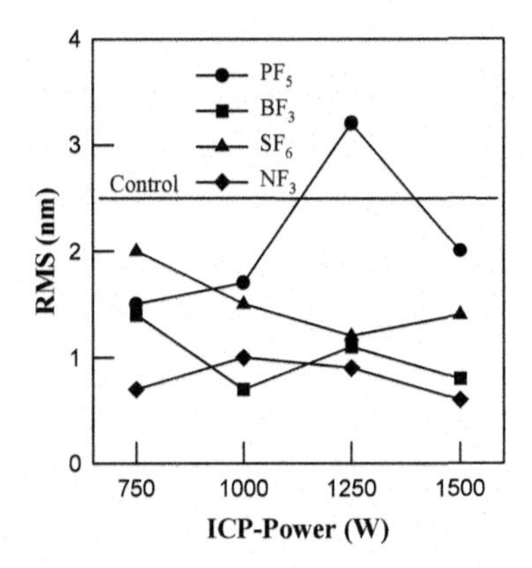

Figure 6. RMS surface roughness of SiC surfaces as a function of ICP source power after etching in pure NF_3, SF_6, PF_5 or BF_3 discharges (250W rf chuck power, 2 mTorr).

5. Mask Materials

Standard conditions of 750W source power 250W rf chuck power and 2mTorr for NF_3 discharges, and addition of O_2 to the chemistry were examined for their effects on etch selectivity of SiC to the different mask materials. Figure 7 shows the etch rates (top) and resultant selectivities for SiC over the masks (bottom) for NF_3/O_2 discharges, as a function of NF_3 percentage of the gas load (15 sccm). The etch rates increase with NF_3 composition for SiC and the mask materials. At high O concentrations there is actually net deposition on Al as it oxidizes, so that the SiC selectivity over Al is infinite.

However the requirement for via hole etching is that the SiC etch rate be > 4000 Å-min^{-1}. Maximum selectivities were > 20 over Ni and ~7 over Al. Note that there were unacceptably low selectivities for SiC over ITO.

Figure 8 shows scanning electron micrographs of features etched ~ 60μm (top) or 100μm (bottom) into SiC substrates. The top micrograph shows the effect of feature diameter in etch depth – the smaller diameter features (~30μm) are shallower by ~15% than the larger openings, which gives the magnitude of the aspect ratio dependent etch rate. The bottom micrograph shows features etched all the way through 100μm thick SiC substrates mounted on sapphire substrates.

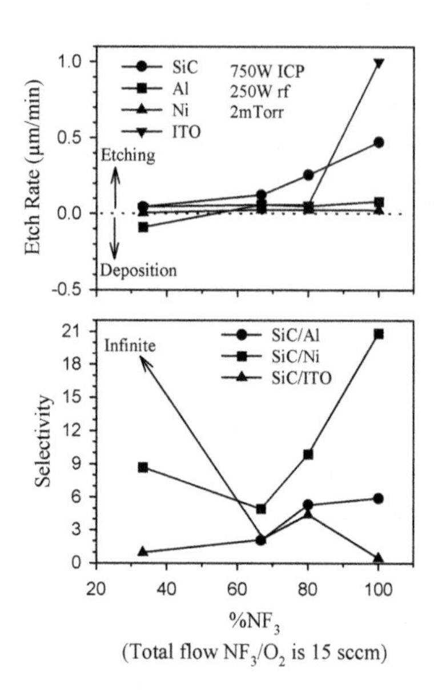

Figure 7. Etch rates (top) and selectivity for SiC over various mask materials (bottom) as a function of % NF_3 in ICP NF_3/O_2 discharges (750W source power, 250W rf chuck power).

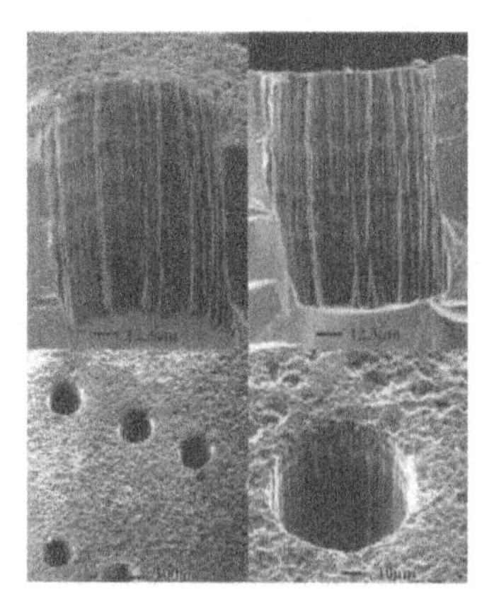

Figure 8. SEM micrographs of deep features etched into SiC substrates using SF_6 discharges.

In situations where only a mesa etch is required, it is desirable that the pattern transfer process not degrade the electrical properties of the SiC. If higher rates are desirable, then the majority of the etching can be performed at higher dc self-biases and this latter parameter can be decreased toward the end of the process.

It is also desirable that there is high selectivity for etching SiC over the mask material (and also the front-side metallization in the case of via holes). Figure 9 shows the dependence of SiC etch rate (top) and selectivity for SiC over Al (bottom) as a function of O_2 percentage (by flow) in 500W source power, 150W rf chuck power, SF_6/O_2 discharges. Note that the SiC etch rate initially increases as O_2 is added to the SF_6. This is probably due to the increase of atomic fluorine neutrals present at low O_2 percentages, a feature well established for CF_4/O_2 and SF_6/O_2 plasma chemistries. The etch rate falls off at higher O_2 percentages because atomic oxygen does not appear to play an active or direct role in etching of SiC. However, the etch selectivity over Al increases rapidly with O_2 addition, since the Al oxidizes and does not etch beyond ~ 40% O_2 addition to the SF_6.

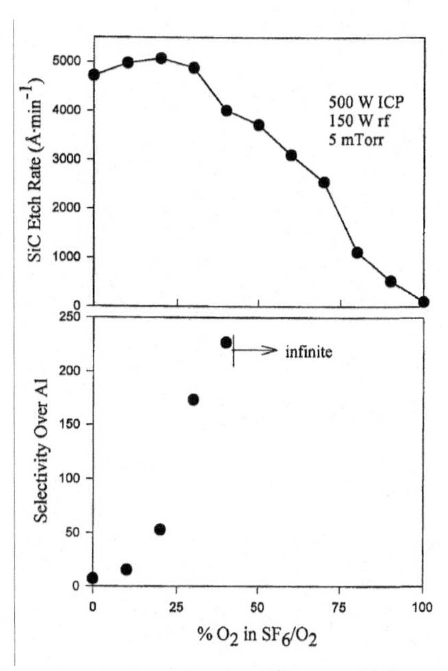

Figure 9. SiC etch rate (top) and selectivity for SiC-over-Al (bottom) as a function of O_2 percentage in SF_6/O_2 discharges (500W source power, 150W rf chuck power, 5mTorr).

The fact that ion energy is a key factor in determining the SiC etch rate is evident in the data of Figure 10. At fixed source power, the incident ion energy is controlled by the sum of this dc self-bias and the plasma potential (-20-25V in this particular tool). The etch rates are always slightly higher with SF_6/O_2 (25% O_2 by flow in this case) compared to pure SF_6 and the rates begin to saturate beyond ~350V where Si-C bond breaking is no longer the limiting step. We should also mention that passing hot gases such as Cl_2, F_2, H_2 and HCl over SiC at high temperatures (> 1200°C) will etch the surface and this process is often employed to clean SiC substrates prior to epitaxial growth.

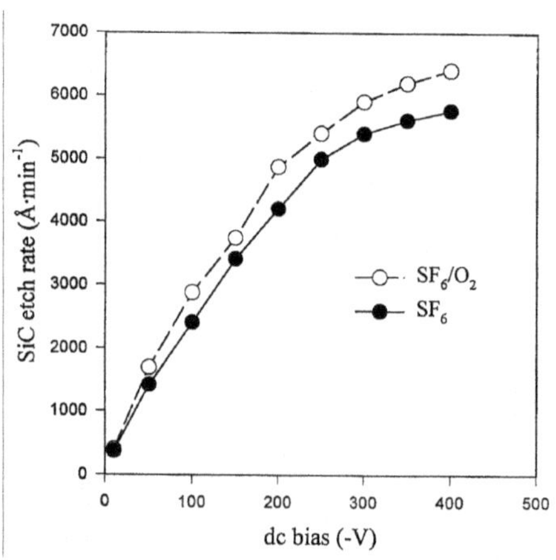

Figure 10. SiC etch rate versus dc self-bias in SF_6 of SF_6/O_2 (25% O_2 by flow rate), ICP discharges.

6. Recent Developments and Future Trends

It has also been recently shown that use of UV illumination during plasma etching in Cl_2-based gas chemistries can enhance the etch rates of SiC, probably through photo-excitation of the chlorinated etch products. This process does not produce any increase in etch rates with F_2-based gas chemistries, because the etch products are already quite volatile.

The achievement of high etch rates for SiC in the various high density plasma sources has now placed the emphasis on developing mask materials that can withstand long plasma exposures, such as needed during via hole formation. The Al masks described earlier work well most of the time provided the residual stress in the metal is minimized. However to pattern smaller features, one would ideally like to avoid thick metal masks and use more convenient materials such as photoresists or dielectrics. Unfortunately these materials etch more rapidly than SiC in F_2-based plasmas, limiting their application to the etching of shallow features.

Since it is clear that more dissociated plasmas with separate control of ion energy produce the fastest etch rates for SiC, it is likely that even higher source powers will be employed in future. Most of the etching to date has been carried out at source powers $\leq$ 1500 W, but reactors are available with powers of 3-5 kW. The higher ion fluxes in these systems will place even greater demands on the durability of mask materials.

An ICP process based on SF_6 or NF_3 provides practical etch rates for deep patterning of SiC. The use of the former gas is probably favored due to its much lower cost and the simpler, less expensive regulators required. Other F_2-based plasma chemistries involving PF_5 or BF_3 do not produce adequate SiC etch rates. Through-wafer vias have been demonstrated using the ICP SF_6 process, as well as low damage conditions for etching of mesas. More conventional RIE techniques can also be employed in most situation in device processing, but suffer from lower etch rates and poorer surface morphologies.

7. Summary

The etching of very deep features for MEMS structures in SiC substrates in practical time frames appears feasible, using the combination of ICP NF_3 or SF_6 discharges and thick metal masks. Addition of O_2 to the plasma chemistry increases etch selectivity for SiC over Al under some conditions, due to oxidation of the Al. In contrast, with very low O_2 concentrations, Ni shows better mask performance. The selectivity for SiC over Ni under this condition is up to 20. Etch rates in excess of 8,000 Å-mim^{-1} have been achieved for 5x5 mm^2 samples of SiC, with ~50% of this area exposed to the plasma. The etch rates for larger samples will be less due to loading effects. Based on our experience with other materials, the fall-off is likely to be of the order of 20-30% when scaling to 3" diameter wafers with ~10% of the area exposed to the plasma. The SiC etch rates with PF_5 and BF_3 are much lower than with NF_3 and SF_6, which is a result of their lower dissociation efficiency in the ICP source.

The main results of etch rate enhancement with UV illumination may be summarized as follows:

SiC etch rates in ICP Cl_2/Ar discharges can be increased by UV illumination.

The mechanism for the enhancement is still the subject of investigation. By analogy with past results on Cl_2/Ar etching of Cu, it is possible that the UV light is absorbed by $SiCl_x$ and CCl_x species on the SiC surface, promoting more complete coordination, i.e. $x \rightarrow 4$. We rule out any change in chlorine radical concentration in the plasma because optical emission spectroscopy showed that the intensity of these emission lines was unchanged by the UV illumination. Similarly we do not believe that sample heating explains the results, because of the absence of any enhancement with SF_6/Ar discharges and the stability of the photoresist masks. The surface morphologies were similar to those obtained without UV illumination.

There was no effect on SiC etch rates in SF_6/Ar ICP discharges. This may be because the etch products are already quite volatile in this case, and desorption of these species is not the rate-limiting step. Rather it is likely that either the initial bond-breaking that must precede etch product formation or the supply of reactive fluorine radicals to the SiC surface are the limiting factors, depending in the exact plasma conditions. The etched surfaces were in general smoother with UV illumination, which may be a result of more uniform desorption of the etch products. In this case the effect of the UV photons may be to increase surface mobility of the adsorbed fluorine.

REFERENCES

1. CRC Handbook of Metal Etchants, ed. Walker, P. and Tarn, W.H. (CRC Press, Boca Raton, FL. 1991), pp.1092-1101.
2. Harris, G.L., in *Properties of SiC*, Ed. G.L. Harris, EMIS Data Review No. 13 9INSPEC, London, UK, 19950, pp.134-135.
3. Buckley, D., *J. Vac. Sci. Technol.*, **A3,** (1985), p.762.
4. Chu, T.L. and Campbell, R.B., *J. Electrochem. Soc.*, **112,** (1965), p.955.
5. Liebmann, W.K., *J. Electrochem. Soc.*, **12,** (1964), p.885.
6. Wolley, E.D., *J. Appl. Phys.*, **37,** (1966), p.1588.
7. Griffith, L.B., *J. Phys. Chem.* Sol., **27,** (1966), p.257.
8. Brander, R.W., *J. Electrochem. Soc.*, **12,** (1964), pp.881.
9. Nordquist, P., Lessoff, H., Gorman, R.J., and Gripe, M.L., Springer. *Proc. Phys.*, **43,** (1989) pp.119.
10. Pirouz, P., Chorey, C.M. and Powell, J.A., *Appl. Phys. Lett.* **50,** (1987), p.221.
11. Jepps, M.W. and Page, T.F., J. *Microscopy,* **124,** (1981), p.227.
12. Shor, J.S. in ref. [2], pp.141-149
13. Yih, P.H. and Steckl, A.J., *J. Electrochem. Soc.*, **142,** (1996), p.312.
14. Wu, J., Parsons, J.D. and Evans, D.R., *J. Electrochem. Soc.*, **142,** (1995), p.669.
15. Casady, J., Luckowski, E.D., Bozack, M., Sheridan, B., Johnson, R.W. and Williams, J.R., *J. Electrochem. Soc.*, **143,** (1996), p.1750.
16. Flemish, J.R., Xie, K. and Zhao, J., *Appl. Phys. Lett.*, **64,** (1994), p.2315.
17. Flemish, J.R. and Xie, K., *J. Electrochem. Soc.*, **143,** (1996), p.2620.
18. Hong, J., Shul, R.J., Zhang, L., Lester, L.F., Cho, H., Hahn, Y.B., Hays, D.C., Jung, K.B., Pearton, S.J., Zetterling, C.-M. and Ostling, M., *J. Electron. Mat.*, **28,** (1999), p.196.
19. Lanois, F., Lassagne, P., Planson, D. and Locatelli, M.L., *Appl. Phys. Lett.*, **69,** (1996), p.236.
20. Wang, J.J., Lambers, E.S., Pearton, S.J., Ostling, M., Zetterling, C.-M., Grow, J.M. and Ren, F., *Solid State Electron.*, **42,** (1998), p.743.
21. Wang, J.J., Lambers, E.S., Pearton, S.J., Ostling, M., Zetterling, C.-M., Grow, J.M., Ren, F. and Shul, R.J., *J. Vac. Sci. Technol.*, **A16,** (1998), p.2204.
22. Leerungnawarat, P., Hays, D.C., Cho, H., Pearton, S.J., Strong, R.M., Zetterling, C.-M. and Ostling, M., *J. Vac. Sci. Technol.*, **B17,** (1999), p.2050.

23. Khan, F.A. and Adesida, I., *Appl. Phys. Lett.*, **75,** (1999), p.2268.
24. Chabert, P., Proust, N., Perrin, J. and Boswell, R.W., *Appl. Phys. Lett.,* **76,** (2000), p.2310.
25. Weitzel, C.E., Palmour, J.W., Carter, C.H., Moore, Jr.K., Nordquist, K.J., Allen, S., Thero, C. and Bhatanagar M., *IEEE Trans. Electron. Dev.,* **43,** (1996), p.1732
26. Xie, K.Z., Zhao, J.H., Flemish, J.R., Burke, T., Buchwald, W.R., Lorenzo, G. and Singh, H., *IEEE Electron. Dev. Lett.,* **17,** (1996), p.142.
27. Baliga, B.J., *IEEE Trans. Electron. Dev.,* **43,** (1996), p.1717.
28. Agarwal, A.K., Augustine, G., Balakrishna, V., Brandt, C.D., Burke, A.A., Chen, L.S., Clarke, R.C., Esker, P.M., Hobgood, H.M., Hopkins, R.H., Morse, A.W., Rowland, L.B., Seshadri, S., Siergiej, R.R., Smith, Jr T.J and Siriam, S., *Tech. Dig. Inst. Electron. Dev. Meeting,* 1996, pp.9.1.1-9.1.6.
29. Brown, D.M., Downey, E., Ghezzo, M., Kretchmer, J., Krishnamurthy, V., Hennessy, W. and Michon, G., *Solid State Electron.,* **39,** (1996), p.1531.
30. Chelnokov, V.E., *Mat. Sci. Eng.,* **B11,** (1992), p.103.
31. Chow, T.P. and Ghezzo, M., *Mat. Res. Soc. Symp. Proc.*, **423**, (1996), p.9.
32. Siriam, S., Clarke, R.C., Burk, A.A., Hobgood, H.M. Jr., McMullin, P.G., Orphanos, P.A., Siergiej, R.R., Smith, T.J., Brandt, C.D., Driver, M.C. and Hopkins, R.H., *IEEE Electron. Dev. Lett.,* **15,** (1994), p.458.
33. Moore, K.E., Weitzel, C.E., Nordquist, K.J., Pond III, L.L., Palmour, J.W., Allen, S. and Carter, Jr. C.H., *IEEE Electron. Dev. Lett.,* **18,** (1997), p.69.
34. Palmour, J.W., Edmond, J.A., Kong, H.S. and Carter, Jr. C.H., *Physica B,* **185,** (1993), p.461.
35. Casady, J.B., Sheridan, D.C., Dillard, W.C. and Johnson, R.W., *Mat. Res. Soc. Symp. Proc.,* **423,** (1996), p.105.
36. Sheppard, S.T., Melloch, M.R. and Cooper, J.A., *IEEE Trans. Electron. Dev.,* **41,** (1994), p.1257.
37. Casady, J.B., Cressler, J.D., Dillard, W.C., Johnson, R.W., Agarwal, A.K. and Siergiej R.R., *Solid State Electron.,* **39,** (1996), p.777.
38. Shenoy, J.N., Cooper, Jr. J.A. and Melloch, M.R., *IEEE Electron. Dev. Lett.,* **18,** (1997), p.93.
39. Agarwal, A.K., Siergiej, R.R., White, M.H., McMullin, P.G., Burk, A.A., Rowland, L.B., Brandt, C.D. and Hopkins, R.H., *Mat. Res. Soc. Symp. Proc.,* **423,** (1996), p.87.
40. Neudeck, P.G., Larkin, D.J., Salupo, C.S., Powell, J.A. and Matus, L.G., *Inst. Phys. Conf. Ser.,* **137,** (1994), p.475.
41. Kordina, O., Bergman, J.P., Henry, A., Fanzen, E., Savage, S., Andre, J., Ramberg, L.P., Lindfelt, U., Hermanasson, W. and Bergman, K., *Appl. Phys. Lett.,* **67,** (1995), p.1561.
42. Casady, J.B. and Johnson, R.W., *Solid State Electron.,* **39,** (1996), p.1409.
43. Yih, P.H. and Steckl, A.J., *J. Electrochem. Soc.,* **140,** (1993), p.1813.
44. Luther, B.P., Ruzyllo, J. and Miller, D.L., *Appl. Phys. Lett.,* **63,** (1993), p.171.
45. Wu, J., Parsons, J.D. and Evans, D.R., *J. Electrochem. Soc.,* **142,** (1995), p.669.
46. Lanois, F., Lassagne, P., Planson, D. and Locatelli, M.L., *Appl. Phys. Lett.,* **69,** (1996), p.236.

47. Flemish, J.R., Xie, K. and Zhao, J., *Appl. Phys. Lett.*, **64,** (1994), p.2315.
48. Flemish, J.R. and Xie, K., *J. Electrochem. Soc.*, **143,** (1996), p.2620.
49. McDaniel, G.F., Lee, J.W., Lambers, E.S., Pearton, S.J., Holloway, P.H., Ren, F., Grow, J.M., Bhaskaran, M. and Wilson, R.G., *J. Vac. Sci. Technol.*, **A14,** (1997), p.885.
50. Flemish, J.R., Xie, K. and McLane, G.F., *Mat. Res. Soc. Symp. Proc.*, **421,** (1996), p.153.
51. Wang, J.J., Lambers, E.S., Pearton, S.J., Ostling, M., Zetterling, C.-M., Grow, J.M. and Ren, F., *Solid State Electron.*, **42,** (1998), p.743.
52. Wang, J.J., Lambers, E.S., Pearton, S.J., Ostling, M., Zetterling, C.-M., Grow, J.M., Ren, F. and Shul, R.J., *J. Vac. Sci. Technol.*, **A16,** (1998), p.2204.
53. Wang, J.J., Lambers, E.S., Pearton, S.J., Ostling, M., Zetterling, C.-M., Grow, J.M., Ren, F. and Shul, R.J., *Solid State Electron.*, **42,** (1998), p.2283.
54. Sugano, T., *Applications of Plasma Processes to VLSI Technology*, (John Wiley and Son Publishers, 1985).
55. Manos, D.M. and Flamm, D.L., *Plasma Etching: An Introduction*, (Academic Press, Boston, 1989).
56. Flemish, J.R., *Dry Etching of SiC. In Proc. Wide Bandgap Semic.*, ed. S.J. Pearton, (William Andrew Publ, Park Ridge, New Jersey, 1999).
57. Muetterties, E.L., *The Chemistry of Boron and its Compounds,* (Wiley and Sons, New York, 1967).
58. Cornbridge, D.E.C., *Phosphorus: An Outline of its Chemistry, Biochemistry and Technology,* (Elsevier, New York, 1978).
59. Emelius, H.J., *The Chemistry of Fluorine and its Compounds,* (Academic Press, New York, 1969).
60. Stacey, M., Tatlow, J.C. and Sharpe, A.G., *Advances in Fluorine Chemistry*, Vol.3, (Butterworths, Washington DC, 1961).

CHAPTER 5

DESIGN, PERFORMANCE AND APPLICATIONS OF SIC MEMS

Stefan Zappe
Edward L. Ginzton Laboratory, Stanford University
450 Via Palou, Stanford, CA 94305-4085, USA
E-mail: zappe@stanford.edu

Micro-electro-mechanical systems (MEMS) based on silicon carbide (SiC) have been emerging over the past fifteen years. In the first part of this chapter the material properties of SiC, applications of SiC MEMS, SiC-based material systems and recent achievements in SiC processing technologies are briefly reviewed. The main part categorizes SiC MEMS devices presented so far, such as pressure sensors, accelerometers, motors, resonators and gas sensors. Their characteristics and performances are summarized. Finally, an evaluation of the state-of-the-art of SiC MEMS is given and further developments are discussed that will be necessary to let SiC MEMS devices reach market maturity.

1. Introduction

The field of MEMS was born in the 1970s when researchers began to investigate more extensively non-electronic applications of the silicon technologies (initially developed for the fabrication of microelectronic circuits).

Since the invention of the first transistor in 1948 (by Bardeen and Brattain under Shockley[1]) and the presentation of the first integrated circuit in 1961, the increasing quality of basic materials, e.g. silicon, and the steady development and improvement of the silicon technologies have enabled the production of integrated circuits, memory chips and microprocessors with an ever-increasing performance.

The establishment of special processes, e.g. anisotropic wet etching of silicon, have enhanced the silicon technologies and have made the fabrication of three-dimensional microstructures possible. In 1982 Petersen reflects this development and describes the role of "Silicon as a Mechanical Material".[2]

The advantages of MEMS products over conventionally fabricated sensor and actuator systems are compelling. Batch processing of wafers and the small size of MEMS chips enable the mass-fabrication of low-cost microdevices. The monolithic integration of electronic circuits and mechanical structures is possible. MEMS technologies provide enormous flexibility to adapt a system design to the requirements of a specific application. This short list already indicates the unique technological and economical potential of MEMS.

Latest efforts in MEMS development aim to explore new material systems, such as silicon carbide, that are suited to serve fields of applications that cannot be covered satisfyingly by silicon.

It was the success of SiC-based electronic and optoelectronic products over the past three decades that lead to a better availability of SiC substrates, an increasing substrate size, improving material quality at decreasing cost and the development of special SiC micromachining processes. This progress has enabled research activities in the field of SiC MEMS as well. Since the early 1990s, SiC has been investigated as a promising material system especially for MEMS devices that are operated in harsh environments under extreme conditions.

And while Petersen looks back in 1995 at more than a decade of commercially available silicon-based MEMS products and strives to identify new trends by asking "MEMS – What Lies Ahead ?",[3] it is Kroetz *et al.* who partially answer the question by emphasizing at the same time and place the importance of "Silicon Carbide as a Mechanical Material".[4]

The following sub-chapters offer brief introductions to the material properties of silicon carbide, fields of applications especially suited for SiC MEMS, available SiC-based material systems and SiC processing technologies that have been developed to create three-dimensional structures and MEMS devices.

1.1. Properties of SiC

Silicon carbide is a compound semiconductor that exists in different polytypes. Usually the Ramsdell notation (e.g. "3C") is used to describe a specific polytype. The number in this notation stands for the number of silicon/carbon double layers that it takes to build up a unit cell of the crystal, while the letter denotes the crystal structure (c-cubic, h-hexagonal, r-rhombohedral). 4H-SiC and 6H-SiC are the most important polytypes for electronic purposes. For SiC MEMS devices, 3C-SiC in addition to 6H-SiC are nowadays preferentially used. The main material properties for the polytypes 3C-SiC and 6H-SiC in comparison to silicon are listed in the appendix in Table 2.

Silicon carbide is a wide bandgap semiconductor with a two to three times wider bandgap than silicon. It has a more than ten times higher breakdown voltage. Its Knoop hardness is about three times as high. Silicon carbide has a higher Young's modulus, for example the elastic coefficient c_{11} of 3C-SiC is more than twice as high as the equivalent value for silicon. Silicon carbide has approximately a three times higher heat conductivity. Unlike silicon, silicon carbide does not melt at 1410°C but sublimes at temperatures of more than 3000°C.

The high Si-C bonding energy makes silicon carbide a very inert material, hard to oxidize and resistant against chemical attacks and radiation damage. At the same time, the formation of a stable silicon dioxide is still possible, a major advantage for example over GaAs.

Silicon carbide exhibits a reasonably high piezoresistive effect that strongly depends on crystal quality, doping level and temperature. While maximum longitudinal gauge factors of approx. $K = -30$ have been reported for 3C-SiC (n-type, <100>-direction) and 6H-SiC (n-type, basal plane) an approximately four times higher gauge factor can be obtained for silicon (n-type, <100>-direction).

1.2. Fields of applications

Based on the material properties discussed above, a number of fields of applications can be identified that significantly benefit from the use of silicon carbide. The following summary covers both pure electronic and

MEMS applications. Numerous introductions to silicon carbide for MEMS and applications of SiC MEMS devices have been given in the past and can provide additional information.[5-14]

Optical Applications Silicon carbide as a wide bandgap semiconductor offers the chance to engineer LED structures that emit light in the blue range – that part of the visible spectrum that cannot be reached by conventional material systems, e.g. based on GaAs.

The first commercially available semiconductor product based on silicon carbide was indeed a blue LED sold by Cree Inc. But since silicon carbide is a semiconductor with an indirect bandgap, the quantum efficiency of SiC LEDs is very low and the optical output power is typically in the μW-range (die size of approx. 200 x 200 μm^2). Nowadays, silicon carbide serves mainly as a substrate for the heteroepitaxial growth of GaN or InGaN-based films. These materials are direct wide bandgap semiconductors and enable much higher quantum efficiencies and a typical optical output power in the mW-range. High-brightness blue and green LEDs are used in traffic lights or full color outdoor displays.

A similar situation is given for blue light emitting lasers where silicon carbide serves as a substrate for the heteroepitaxial growth of the active InGaN-based structure. Applications of blue lasers include the next-generation DVD market.

SiC photodiodes are sensitive in the region between 200 nm–400 nm and thus extend the regime that is covered by silicon photodiodes deeper into the UV range. Applications of UV photodiodes include flame detection and control.

High Temperature Applications Silicon Carbide is well suited for high temperature applications. While silicon deforms plastically under low mechanical stress at temperatures above T = 500°C,[15] silicon carbide is supposed to be mechanically stable even at temperatures well above T = 1000°C. Due to its small bandgap, the intrinsic carrier concentration of silicon reaches a significant level of $n = 10^{16}\ cm^{-3}$ already at approx. T = 400°C. A similar intrinsic carrier concentration e.g. in 3C-SiC will be reached at a much higher temperature of approx. T = 1000°C.

P/n-junctions in silicon exhibit excessive reverse leakage currents starting at T = 125°C. This effect leads for example to a performance degradation of sensors with implanted piezoresistors. In silicon carbide, this effect is shifted to temperatures higher than T = 350°C.

Typical fields of applications for high temperature sensors and electronics are: motor management, oil well logging, industrial process control, turbine or turbo charger engine control, avionics and spacecraft systems. Silicon carbide could make cooling systems for sensors and electronics obsolete and could contribute to a reduced cost, volume and weight of control systems. SiC-based control systems could reduce fuel consumption and thus environmental pollution.

High Power Applications Especially its high breakdown voltage and high heat conductivity make silicon carbide suited for the realization of high power electronic devices. A tenfold increase in blocking voltage compared to conventional silicon devices can be achieved as well as a higher current density in forward mode. Silicon carbide power devices can operate at higher temperatures. They switch faster due to lower parasitic resistances and capacities in consequence of a smaller size compared to silicon devices.

Fields of applications affected by silicon carbide high power devices include electric vehicles, where much higher energy conversion efficiency from electric to kinetic energy can potentially be achieved, and also public electric power distribution. SiC high power devices, e.g. SiC Schottky diodes, are today commercially available.

RF (Radio Frequency) and Microwave Applications Its high breakdown voltage, high heat conductivity and high electron saturation drift velocity also make silicon carbide well suited for RF and microwave devices. Comparable to high power devices, a higher current density and a higher working temperature can be tolerated. Device sizes can be reduced and smaller parasitic capacitances and resistances result in a wider bandwidth at a higher power level compared to Si or GaAs-based RF devices.

Applications for silicon carbide RF and microwave devices are cellular infrastructure, high power radar, wideband military communication and electronic warfare.

SiC RF MESFETs are today commercially available. In June of 2003, the company Cree Inc. announced to offer a silicon carbide monolithic microwave integrated circuit (MMIC) foundry service to U.S. customers based on its 3-inch wafer SiC MESFET production process.

The potential use of silicon carbide is not restricted to electronic RF components. SiC RF MEMS devices, e.g. for cellular communication infrastructure, include switches, inductors and variable capacitors, high Q filters, phase shifters, antennas with reconfigurable radiation characteristics and transmission lines.

Mechanical, Chemical and Radiation Resistance Applications Silicon carbide as a hard and chemically inert material offers advantages over silicon whenever parts of a microsystem are exposed to mechanical friction, aggressive media or high radiation levels. Potential applications cover media compatible sensors, chemically inert microfluidic devices, micromotors and – actuators with enhanced lifetime, and sensors and electronics for high radiation applications, e.g. space missions.

1.3. Available material systems

The performance and reliability of SiC MEMS devices strongly depends on the quality of used material systems. 4H-SiC and 6H-SiC 3-inch wafers with n- and p-type epitaxial layers are today commercially available with a crystal quality suitable for electronic devices.

3C-SiC is usually deposited as a thin film with a thickness of up to several microns, mainly onto silicon and silicon-on-insulator (SOI) substrates. Polycrystalline and amorphous films are used for MEMS devices. When deposited onto crystalline surfaces, e.g. silicon, even single crystalline 3C-SiC can be obtained. However, due to mismatches in thermal expansion coefficient and lattice constant, films are produced with a much higher defect density compared to hexagonal bulk SiC and are less suitable for electronic devices. 3C-SiC material is currently not commercially available. But the company FLX Micro Inc. started to offer the multi-user silicon carbide (MUSIC®) process on a commercial basis. MUSIC® is a multi-user foundry service comparable to the multi-user MEMS processes (MUMPs) service[16] but based on multilayers of polycrystalline SiC. MUSIC® provides the SiC MEMS research community with an affordable access to thin film SiC technologies.

Silicon Carbide Bulk Material The first observation of silicon carbide was made by Berzelius in 1824,[17] while the first process for the industrial fabrication of SiC was developed by Acheson and presented in 1893.[18] The material obtained with this process exhibits a poor crystal quality, suitable only for non-electronic applications (e.g. as an abrasive).

It took several decades until advancements in silicon technologies also gave rise to an interest in silicon carbide for electronic purposes. In the 1950s and 1960s, different NASA and US Air Force research programs pushed the investigation of SiC. In this period, Lely presented in 1955 his method of SiC growth by sublimation of polycrystalline SiC and obtained the first single crystalline material with a reasonable size at least for academic research.[19]

Another period of worldwide extensive research on SiC started at the end of the 1970s. Tairov and Tsvetkov modified the Lely-method and introduced in 1978 the seeded sublimation growth of silicon carbide.[20] This modified process initiated the production of bulk single crystalline material suitable for the fabrication of silicon carbide wafers.

The today's world leader in development, manufacturing and marketing of SiC substrates and SiC-based devices, Cree Inc., Durham, USA, was founded in 1987 by a group of researchers from North Carolina State University. Cree started to produce 4H-SiC and 6H-SiC substrates[21] and offers today wafers with a diameter of three inches, a first four-inch wafer was presented in 1999. Nowadays, 4H-SiC and 6H-SiC wafers with optional epitaxial p- and n-type thin films are offered by a variety of companies.

Heteroepitaxially Grown 3C-SiC Thin Films The heteroepitaxial growth of silicon carbide thin films on different substrates has been investigated but mismatches in thermal expansion coefficients and lattice constants have first led to a poor crystal quality. Matsunami and Nishino *et al.* achieved a major improvement in 1980/81 by introducing the growth of an intermediate buffer layer on top of a silicon substrate to facilitate higher quality heteroepitaxial growth by chemical vapor deposition (CVD).[22,23]

In the beginning, Cree offered 4-inch silicon and later also silicon-on-insulator (SOI) wafers with such heteroepitaxially grown, single crystalline 3C-SiC thin films as a commercial product.[24] The production was stopped in the middle of the 1990s when the material quality of 3C-SiC thin films was still inferior to the quality of hexagonal SiC and the needs of the most interesting market segments (blue LEDs, electronic devices) could not be covered satisfyingly with cubic thin films.

3C-SiC thin films with a crystal quality comparable to what was last offered by Cree were produced in recent years by different research institutions and were mainly used for MEMS applications.[25,5]

The 3C-SiC/SOI Material System While 3C-SiC still plays a minor role for the development of electronic devices, it is promising for high temperature sensors. Such sensors often feature mesa-etched, 3C-SiC thin film piezoresistors that are electrically isolated from the substrate.

But 3C-SiC heteroepitaxially grown directly on silicon is not well suited for high temperature applications. A significant leakage current for example across a Si-SiC pn-heterojunction already occurs at temperatures above $T = 250°C$.

3C-SiC is better deposited onto an SOI wafer (typically a silicon substrate with a several hundred nanometer thick, electrically insulating, buried silicon dioxide layer and a 20-200 nm thin, single crystalline silicon surface layer). Such a silicon carbide-on-insulator (SiCOI) material system has two major advantages: the silicon carbide film provides high temperature capability and is electrically isolated from the silicon substrate up to $T = 600°C$, while the silicon substrate can be easily micromachined using conventional MEMS technology. Up to $T = 600°C$, the thin, low-doped silicon surface layer does not significantly short-circuit the much higher doped 3C-SiC film that is typically several micrometer thick.

Plastic deformation of the silicon substrate under low mechanical stress limits the use of the SiCOI system to approx. $T = 500°C$.[15]

Other SiCOI Material Systems Many other SiCOI material systems have been realized. Direct deposition of 3C-SiC onto sapphire substrates has been demonstrated.[26,27] Polycrystalline 3C-SiC was deposited onto silicon wafers with silicon nitride surface films.[28] Ion beam synthesis (IBS)[29] of silicon carbide films and SiC layer transfer by wafer bonding to oxidized silicon wafers were applied to fabricate SiCOI structures. Direct conversion of polysilicon on top of oxidized wafers into polySiC by IBS resulted in the direct fabrication of a polySiCOI system.[30] Single crystalline 3C-SiC films were transferred from silicon substrates to oxidized silicon wafers by bonding and etchback steps.[31,32] High quality single crystalline 6H-SiC thin films were transferred to oxidized silicon substrates via the Smart Cut® process.[33,34]

The MUSIC® Process The lack of commercially available silicon and SOI substrates with heteroepitaxially grown high quality 3C-SiC thin films has slowed down research efforts related to SiC MEMS during recent years.

This situation has changed with the announcement of the company FLX Micro Inc. at the end of 2002 to make the MUSIC® process commercially available. FLX Micro gained access to the necessary silicon carbide technologies through an exclusive license agreement with Case Western Reserve University where extensive research efforts by Zorman and Mehregany enabled this development. Process details are described e.g. by Song *et al.*[35] Further descriptions and a summary of design rules are available on the company's website.

In short, MUSIC® is a multi-user surface micromachining process that utilizes polycrystalline silicon carbide structural layers. The MUSIC® process provides low-cost access to MEMS prototyping by combining multiple chip designs onto a single large-area substrate. A die size of 1 cm x 1 cm is offered to customers. MUSIC® is an eight-mask surface micromachining process. Micromolding and polishing steps are used to create planarized structural layers. At the end of the fabrication process the sacrificial polysilicon and silicon dioxide molds are dissolved to release fabricated SiC MEMS devices.

The commercial availability of this service is well suited to accelerate SiC MEMS research.

1.4. SiC MEMS fabrication technologies

Powerful micromachining technologies are as essential as the availability of high quality material systems for the successful development and reliable performance of SiC MEMS devices. This sub-chapter will briefly discuss the latest advances in SiC MEMS fabrication technologies.

Whenever 3C-SiC thin film structures on the surface of a silicon or SOI substrate are needed, one may choose selective deposition of SiC onto pre-structured wafers.[36,37] The choice of suitable process parameters leads to the deposition of crystalline SiC on silicon islands, while no SiC is deposited onto surrounding silicon dioxide or silicon nitride surfaces. Standard silicon technologies can be applied to the silicon substrate during remaining fabrication steps.

If a selective deposition process is not available or applicable, already standard SF_6/O_2 plasma etch chemistry has been proven to etch successfully several micrometer thick continuous silicon carbide films.[38] A one-step SF_6/O_2 inductively coupled plasma (ICP) etching process for the fabrication of SiC MEMS devices has been developed recently.[39] More dry etching processes have been presented in the past,[40,41] exhibiting satisfying selectivity, uniformity, and anisotropy. Anisotropic processes with etch rates of 1 μm / min and higher do not only allow to pattern thin films but enable nowadays whole wafer silicon carbide bulk micromaching.[40,41,42]

Alternatives to deep reactive ion etching are (photo-)electrochemical wet etching processes.[40,43,44] SiC is first anodized to form a deep porous layer, and this layer is subsequently removed by thermal oxidation followed by a dip in hydrofluoric acid (HF). Anodization rates for example of 400 nm/min (n-type 6H-SiC), 2.2 μm/min (p-type 6H-SiC) and up to 100 μm/min for n-type 3C-SiC were reported.[43,45] A controlled etch depth and the selective etching of multilayer films can be achieved by applying dopant-selective etch stop techniques.[46,47]

Laser ablation has been used to structure silicon carbide films with thicknesses over 1 μm.[48] Focused ion beam sputtering of 6H-SiC for the fabrication of holes was succesfully investigated.[49]

A surface micromachining process has been developed that uses the deposition of polycrystalline silicon carbide into silicon dioxide and polysilicon molds, polishing of the SiC film on top of the mold structure and release through selective etching.[50] A similar process for bulk micromachining has been demonstrated that uses deep reactive ion etching (DRIE) of a silicon mold, the deposition of polycrystalline SiC into the mold, subsequent polishing and release of the SiC structure by dissolving the silicon mold.[51]

The successful chemical mechanical polishing (CMP) of 3C-SiC films with removal rates up to 0.58 μm/h and a remaining surface roughness of 1.5 nm was presented.[52] 3C-SiC films were successfully transferred to different oxidized silicon substrates through polysilicon deposition onto the SiC film, polishing, silicon to oxide bonding and removal of the initial substrate.[31,32] The Smart Cut® [53] process was successfully applied to transfer thin 6H-SiC films to oxidized silicon and polycrystalline silicon carbide wafers.[33,34] SiC-to-SiC whole wafer bonding was recently demonstrated at temperatures of $T \geq 800°C$ and under a uniaxial mechanical stress of 20 MPa.[54]

The reliable electrical interconnection to SiC MEMS structures is important. Ohmic and rectifying electrical contacts to n- and p-type SiC have been realized.[55] For example the system W/WC/TaC/SiC exhibited excellent stability even when annealed at $T = 600°C$ for 1000 hours.[56]

High temperatures compatible die attach schemes were developed using precious metal based thick-film material and aluminum nitride and alumina ceramic substrates.[57] Attached test dies using those schemes survived both electronically and mechanically performance and stability tests at 500 °C in an oxidizing environment of air for 550 hours.

A nickel wire bonding technique was developed and applied to Ni contact pads on 3C-SiC films using conventional tools and wire diameters and reliable operation up to $T = 550°C$ was demonstrated.[58]

2. SiC MEMS Devices

The brief summaries of available SiC-based material systems and micromachining technologies indicate the tremendous progress of the past several years that gave rise to the development and demonstration of numerous SiC MEMS devices. This main chapter categorizes devices presented so far and describes in more detail their performances and characteristics. Pure electronic devices (e.g. power and RF transistors, integrated and hybrid circuits) are not discussed, while field effect devices used for gas sensing are considered as MEMS and are treated below. Optoelectronic and radiation detection devices are briefly covered for their potential use in MEMS systems as well. Various books about MEMS can provide additional information on the theory, design and fabrication of MEMS devices in general.[59-65]

2.1. Test structures

The successful design, simulation, fabrication and operation of SiC MEMS require the knowledge of material properties. These properties strongly depend on the method used to synthesize SiC. Test structures and methods aimed to determine Young's modulus, residual stress, thermal conductivity and piezoresistive properties of SiC thin films are discussed in this sub-chapter.

2.1.1. Young's modulus and residual stress

Many of the test structures developed so far aimed to investigate the mechanical properties of deposited silicon carbide thin films such as Young's modulus and residual stress.

Mehregany *et al.* presented in 1997 measurements on a 5 μm thick single crystalline 3C-SiC film, deposited via CVD on a silicon wafer. Residual stress and biaxial modulus were determined by load–deflection measurements of suspended 3C-SiC diaphragms. The film's residual stress was tensile with an average of 212 MPa, while the in-plane biaxial modulus averaged 441 GPa.[66]

Serre *et al.* measured in 1999 the mechanical properties of SiC films by a beam bending method using an atomic force microscope.[67] Crystalline 300 nm thick 3C-SiC layers were used, obtained by high temperature multiple step carbon implantation at T = 500°C into silicon and subsequent annealing.[29]

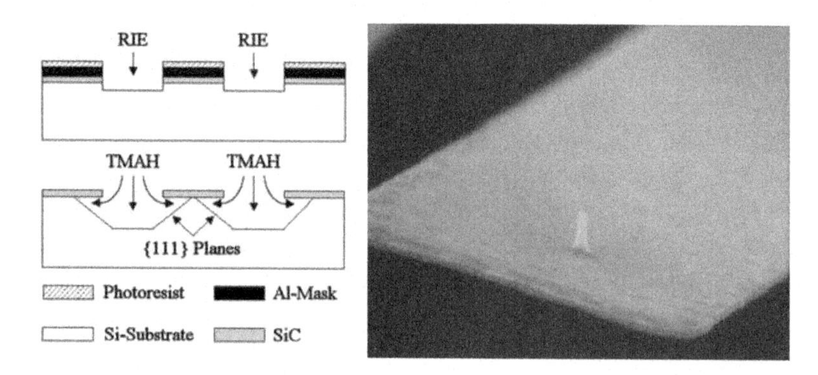

Figure 1. Left: fabrication process of a 3C-SiC cantilever for Young's modulus measurements. Right: SEM image of a 300 nm thick 3C-SiC cantilever with an electron beam-deposited marker.[68] SiC was formed by ion implantation of carbon into silicon.[30]

The low residual stress level in the layers along with the high stiffness and excellent etch-stop properties of SiC allowed the fabrication of free standing cantilevers using standard silicon bulk micromachining techniques (Figure 1). Values for the Young's modulus in the range of 447 - 487 GPa were measured.

2.1.2. Piezoresistive effect

Several groups used deflection methods applied to cantilevers with patterned homo- and heteroepitaxially grown silicon carbide thin films to determine SiC gauge factors. Measurements on material with a crystal qualities that still can be considered state-of-the-art are discussed below.

The dependence of the piezoresistive effect on the crystal orientation in 3C-SiC and silicon is very similar.[69,70] In n-type 3C-SiC, a maximum effect occurs in <100>-direction, while in p-type 3C-SiC the maximum effect is measured in <110>-direction. The piezoresistive effect of 6H-SiC as a hexahedral crystal is isotropic within the basal plane[71] and gauge factors of both n- and p-type material have been investigated.

In 1993 Shor *et al.* reported the longitudinal (transverse) gauge factor of n-type 3C-SiC in <100>-direction to be $K = -31.8$ ($K = +19$) for a film with a resistivity of $\rho = 0.7\ \Omega$ cm. The longitudinal gauge factor decreased with decreasing resistivity to $K = -12.7$ ($\rho = 0.002\ \Omega$ cm).[70] The gauge factors decreased with temperature to approximately half their room-temperature values at 450°C. Reichert *et al.* reported in 1996 a room temperature gauge factor of $K = +18.7$ in <110>-direction for p-type 3C-SiC with a doping level of $N_A=2\cdot10^{18}$ cm^{-3}.[72]

In 1994 Shor *et al.* presented measurements on n-type 6H-SiC resistors.[73] Within the basal plane a longitudinal gauge factor of $K = -29.4$ (doping level of $N_d = 3.3 \times 10^{18}$ cm^{-3}) was measured with a decrease by approx. 43% at T = 250°C. For the same film a transverse gauge factor of $K = +20$ was measured, dropping with temperature to $K = +12$ at T = 250°C.

Highly doped n- and p-type 6H-SiC films with doping levels of $N_a = N_d = 2.2 \times 10^{19}$ cm^{-3} were investigated by Okojie *et al.* in 1998.[74] A longitudinal gauge factor of $K = -22$ (n-type) was measured, decreasing to 52% at 250°C. The transverse gauge factor decreased from $K = +6$ (n-type) to $K = +3.3$ at 250°C. The longitudinal gauge factor of a p-type film decreased from $K = +27$ to 55% of this value at 250°C.

2.1.3. Thermal conductivity

Jansen *et al.* reported in 1998 on thermal conductivity measurements using microstructures with LPCVD-grown 3C-SiC thin film membranes and integrated aluminum heater and thermistors. A maximum heat conductivity of $k_T \approx 73$W/mK was measured at T = 80 °C.[75] Irace and Sarro presented in 1999 measurements using microstructures with double-layer membranes consisting of a supporting LPCVD silicon nitride film and an amorphous PECVD silicon carbide film.[76] Thermal properties were investigated with integrated polysilicon and aluminum heater and thermocouples. The extracted thermal conductivity of $k_T \approx 150$W/mK for the PECVD SiC film is approx. three times lower than a value obtained for high crystal quality, bulk 3C-SiC material.[77]

2.2. Pressure sensors

The pressure sensor market is probably the largest and most important for MEMS products in general. High temperature applications especially have led to the development of numerous silicon carbide-based prototypes as well. The main advantage of SiC-based pressure sensors is their ability to withstand high operation temperatures without cooling system. Hence, they can be smaller in size and are more cost effective to fabricate and operate than silicon-based devices.

Ziermann and von Berg *et al.* successfully developed and demonstrated in the late 1990s a silicon carbide-based, membrane-type piezoresistive sensor for pressure measurements inside the combustion chamber of a gasoline engine.[78-84] An on-chip integrated thermistor facilitated temperature compensation.[84] Inserted like a spark plug, such a packaged sensor can monitor directly the pressure during combustion cycles. It helps to recognize failure, to optimize combustion parameters and to increase fuel combustion efficiency.

Figure 2 depicts the sensor chip that is based on the 3C-SiC/SOI material system and shows also a scheme of the packaging. The 3C-SiC piezoresistors and the silicon substrate were both dry-etched using an SF_6/O_2 plasma. The chips were successfully tested up to $T = 400°C$. The packaged sensor was installed in a motor block and it reliably recorded pressure peaks at partial and full load. The pressure signal of the SiC-based sensor was almost identical compared to a reference signal obtained with an expensive, water-cooled, piezoelectric quartz reference sensor (Figure 3).

Zappe *et al.* presented at the end of the 1990s high-pressure sensor chips ($P_N = 1000$ bar) suitable for oil well logging applications[85-88] and low-pressure sensor chips ($P_N = 6$ bar) with enhanced pressure overload range for turbine control applications.[86,89] These sensor chips were also based on the 3C-SiC/SOI material system with TiWN/Au metallization. They were successfully tested up to $T = 400°C$, at pressures up to $P = 700$bar and $P = 65$ bar, respectively. Electrical measurements of single resistors at temperatures up to $T = 750°C$ indicated sufficient electrical insulation of piezoresistors and extreme heat-shock survival capability.[84] Oil well logging requires good long-term stabilities and extremely low drifts (< 0.5 bar / month @ 250°C). Efforts were made to improve material stability by reducing the density of defects at the interface between SiC thin film and silicon overlayer through the introduction of stabilizing buried silicon nitride layer under the silicon

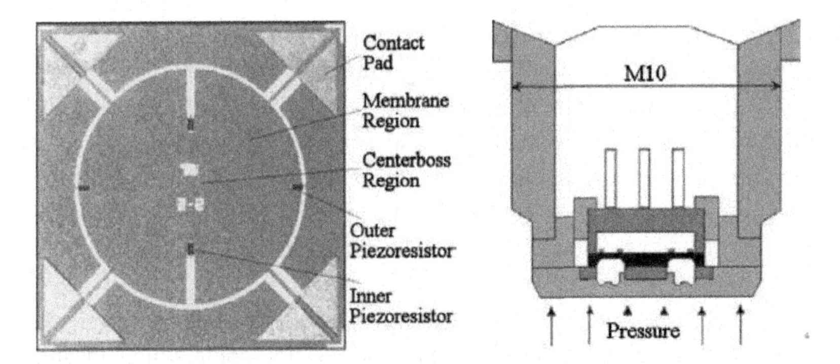

Figure 2. Left: membrane-type sensor chip with mesa-etched 3C-SiC/SOI thin film piezoresistors and TiWN/Au metallization system. Right: packaged sensor chip, sandwiched between two ceramic pieces and clamped between a steel cylinder and a pressure sensing steel membrane.[81]

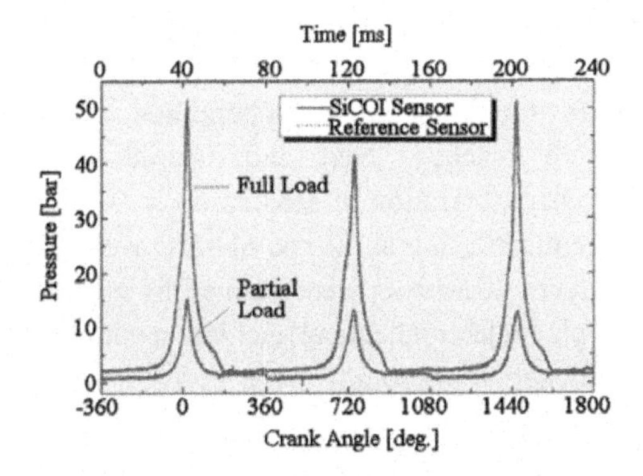

Figure 3. Measurement of the pressure inside a combustion chamber of a gasoline engine vs. time (crank angle). The SiC-based sensor produced a signal that is almost identical to that of a water-cooled, piezoelectric, quartz reference sensor.[81]

overlayer.[90-92] But despite encouraging results, the sensor tests revealed that a major improvement in material quality or a change to bulk silicon carbide sensors is necessary to meet such extreme stability demands.[86]

In 1999, Eickhoff *et al.* developed another membrane-type piezoresistive pressure sensor based on the 3C-SiC/SOI material system.[93] For the first time. the sensing elements were defined through selective deposition of 3C-SiC on a patterned Si/SiO_2 surface. A typical gauge factor of K= -18 at room temperature (K= -10 at 200°C) indicated the successful deposition of single-crystalline 3C-SiC.

Kroetz *et al.* discussed in 1999 in general the application of SiC MEMS in harsh environments and also presented a different type of combustion pressure sensor.[7] In 2001 Wu *et al.* presented another pressure sensor based on 3C-SiC.[94] The material system 3C-SiC/SiO_2/Si was fabricated using bonding and etchback techniques. The devices were characterized up to T = 385°C.

Pakula *et al.* commented in 2003 on the fabrication and characteristics of a surface micromachined, capacitive, CMOS-compatible, absolute pressure sensor for harsh environments.[95] The sensor consisted of 100 circular membranes with a total capacity of 14 pF. PECVD SiC films were used to fabricate the membranes.[96] The polyimide PI2610 served as material for a sacrificial layer. A pressure change from 10 mbar to 1.01 bar resulted in a capacitance change of 0.4 pF.

The successful operation of 6H-SiC-based sensors was reported by Okojie *et al.* Maximum operation temperatures of T = 350°C[97] and T = 500°C[98,99] were reached. Ned and Okojie *et al.* eventually demonstrated the characterization of sensors up to T = 600°C.[100] These sensors chips were made using an n-type 6H-SiC wafer with p-type and n-type epitaxial layers. Resistors were defined by etching of the n-type epilayer using a photoelectrochemical etching method.[101] Membranes with a typical thickness of 40 μm were defined by electrochemical etching from the backside. Ti/TiN/Pt/Au was sputtered as high temperature capable metallization system. Ni/Au-plated Kovar headers and gold wire bonding was used. With a 5V supply voltage a full scale output (FSO) voltage of 66.4mV at room temperature and P = 1000 Psi was measured. A very low hysteresis of 0.7% FSO and a good linearity of –0.9% FSO were obtained. The FSO voltage of the sensor dropped by 62% to 25mV at T = 600°C.

2.3. Accelerometers

Comparable to pressure sensors, there is also a substantial market for accelerometers. A well-known application is the use of accelerometers in car airbag systems. Other applications, especially at elevated temperatures, include vibration control and frequency analysis, e.g. during oil well drilling, motor management and analysis, avionics, spacecraft and missile control.

Okojie and Atwell *et al.* presented in 2002/03 the design, fabrication and test of bulk micromachined, high-g, piezoresistive accelerometers based on 6H-SiC for military and space applications.[102,103] They attempted to meet extreme demands regarding impact (>100,000g), high electromagnetic field (>18T in some railguns) and high temperature (>500°C).

A scanning electron microscope image of the so-called ACLBOS400 configuration and a device cross-section to depict the relative positions of the piezoresistors are shown in Figure 4. The four piezoresistors are placed longitudinally on the narrow beams, each located on the inner edge of the peripheral rigid structure and at the opposite edges of the centered inertial proof mass (boss). Both the front-side and backside were structured by a deep reactive ion etching process.[105] The thickness of diaphragms and cantilevers is 60 μm, while centerboss and wafer are 400 μm thick.

Test results at room temperature for a slightly modified version without slots in the membrane, the so-called PREBOS400 design, are provided in Figure 5. The shown sensor response is compared to that of a commercially available system (ENDEVCO 7270-60 K). The SiC device reasonably replicates the output of the benchmark sensor for the first acceleration peak until a rapid reversal of the signal is encountered approx. at 0.05 ms, most likely due to the asymmetric rectangular sensor design and asymmetric dynamic responses of wide and narrow diaphragm regions. The reversal of the signal was not encountered until the shock test level reached 8,000*g*. In the range below 8,000*g* PREBOS400 matched the benchmark output in frequency and magnitude for the shock pulse. The maximum acceleration applied was 40,000*g*.

Other results of sensors for high-g launch application were presented by Katulka in 2002.[106] Mehregany and Zorman presented in 2001 the feasibility of a surface micromachined, capacitive accelerometer based on a parallel plate approach.[6]

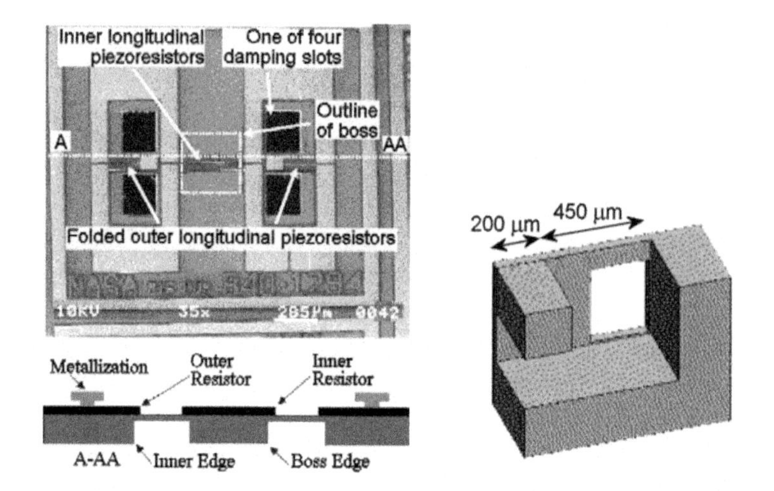

Figure 4. Left, top: SEM image of the ACLBOS400 accelerometer chip, depicting four longitudinal piezoresistors placed on narrow beams, the boss location and four damping slots. Left, bottom: A-AA cross-section indicating the locations of the inner and outer resistors on the narrow beams. Right: scheme of the bottom view at one quarter of the device.[104]

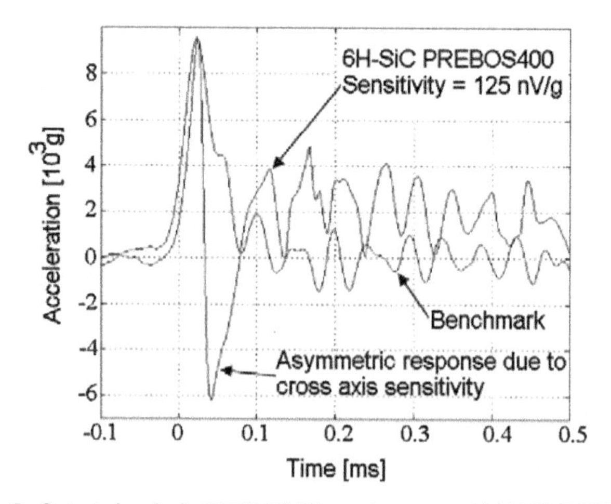

Figure 5. Output signal of a PREBOS400 accelerometer chip (ACLBOS400 design without damping slots) in comparison to the response of a benchmark sensor at 9000*g*. A rapid reversal of the PREBOS400 chip signal occurs at 0.05 ms.[104]

2.4. Resonator structures

Micromachined resonators are emerging as potential on-chip replacements for conventional discrete oscillators and filters in high performance communication transceivers.[107] Other applications include mechanical property testing, pressure sensing and inertial navigation systems.

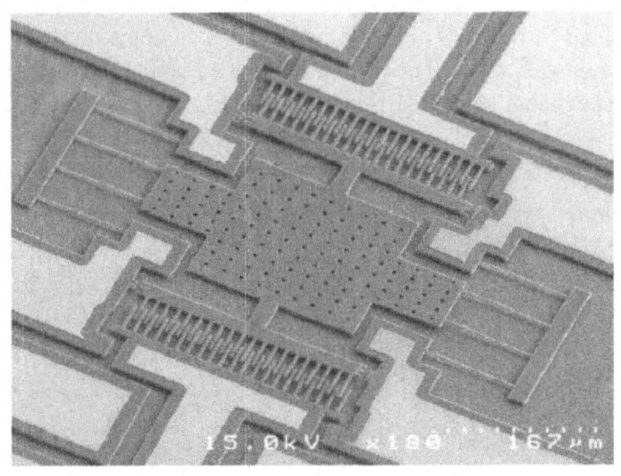

Figure 6. SEM micrograph of a released polySiC lateral resonant device. The suspension beam lengths and widths are nominally 100 µm and 2.5 µm, respectively. Exposed polySiC shows up as dark gray, while the Ni metallization appears light gray.[111]

The modeling of SiC lateral resonant devices was discussed by De Anna *et al.* in 1999/2000.[108,109] The fabrication and characterization of polycrystalline SiC resonators was presented by Roy *et al.* in 2002.[110] A three mask surface micromachining process using silicon dioxide, polysilicon, and nickel (Ni) as the isolation, sacrificial, and contact metallization layers, respectively, was applied to fabricate resonators as shown in Figure 6. The polySiC resonators were packaged for operation in high temperature environments using ceramic-based materials and nickel wirebonding procedures. Device operation was successfully demonstrated over 10^{-5}–760 torr and 22–950°C pressure and temperature ranges, respectively. Quality factors of 100,000 at 10^{-5} torr and resonant frequency drifts of 18 ppm/h under continuous operation were achieved using a scanning electron microscope (SEM) setup. Device resonant frequency varied nonlinearly with increasing operating temperature.

Finite element modeling revealed that this variation resulted from the interplay between the Young's modulus of polySiC and induced stresses, which occur due to mismatch in thermal expansion coefficients of the polySiC film and the underlying silicon substrate.

A summary of resonant frequency, quality factor and Young's modulus for resonators with different geometries is given in Table 1.

Table 1. Resonant frequency f_r, quality factor Q and Young's modulus E of polySiC resonators, measured at $P < 10^{-5}$ torr. Nominal Device Parameters – W: suspension beam width, L: suspension beam length, M_p: shuttle mass, M_t: mass of trusses, and M_b: mass of suspension beams. Errors in f_r and Q arise from uncertainty in resolution of resonator amplitude. Error in E arises from uncertainties in resonator geometry.[112]

Device #	Resonant Frequency, f_r (± 0.01Hz)	Q (±15%)	Young's Modulus, E (±75 GPa)	Nominal Device Parameters W, L (1E-6 m) M_p, M_t, M_b (1E-10 kg)
A1	10060.15	62888	324	2.75, 250 3.62, 0.45, 0.16
A4	13886.41	86738	316	2.75, 200 3.62, 0.45, 0.12
B2	11359.93	107926	414	2.75, 250 3.62, 0.45, 0.16
C1	28920.43	57828	310	2.25, 100 3.62, 0.45, 0.05

2.5. Motors

The fabrication and characterization of a variety of silicon micromotors were published. Applications include microshutters in optical setups.[113] But the surface to volume ratio in such microsystems is very high; wear of mechanical parts plays an important role and leads to short device lifetimes. Silicon carbide as a much harder material compared to silicon enables the fabrication of devices with a prolonged lifetime and an improved reliability.

Yasseen *et al.* presented in 2000 the fabrication and testing of surface micromachined polycrystalline silicon carbide micromotors.[114] A new multilayer fabrication process utilizing low temperature deposition and micromolding techniques was developed to create the desired SiC structural components. Typical operating voltages of salient-pole[12] and wobble micromotors (Figure 7) in room air were 100V and 80V, respectively. Wobble micromotors were tested at room temperature in atmospheres of argon, nitrogen, oxygen, and room air (25% humidity). In addition, micromotors were tested at elevated temperatures and exhibited stable operation up to 500°C.

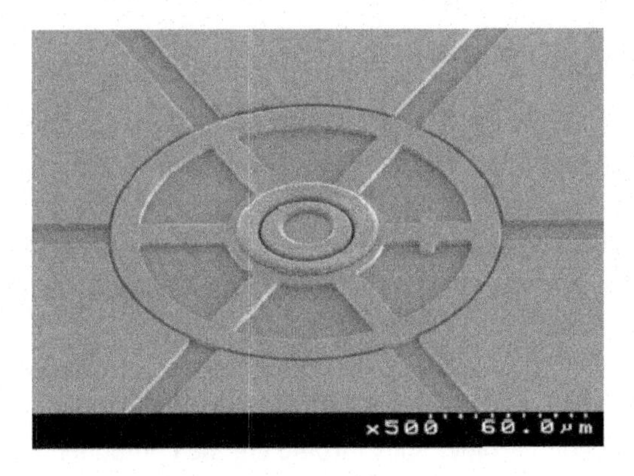

Figure 7. SiC wobble micromotor after dissolving the polysilicon mold.[115]

A parameter that characterizes the function of wobble motor is the so-called gear ratio. The gear ratio is defined as the ratio of the electrical excitation to rotor rotational frequencies. Under ideal conditions, the gear ratio equals the ratio of the bearing radius to the bearing clearance. Ideal conditions imply pure rolling of the rotor on the bearing (i.e. no rotor slip) during operation.

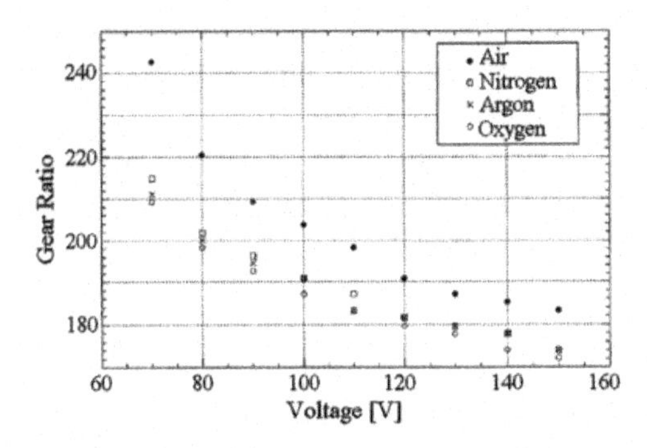

Figure 8. Gear ratio versus excitation voltage for a 150 μm-diameter SiC wobble micromotor with a 2 μm-wide gap tested in air, nitrogen, argon, and oxygen.[115]

As shown in Figure 8, the gear ratio as a function of applied voltage was higher for operation in room air as compared with the other gases, suggesting a relationship between gear ratio and relative humidity.

2.6. Wear resistant devices

Silicon carbide with its high hardness, chemical inertness and ability to survive operation at high temperatures and in harsh environments is well suited to be used as a protective coating for example of micromachined silicon parts. Furthermore it can directly be used as structural material for the fabrication of reliable MEMS parts with an enhanced lifetime.

The use of micromachined fuel atomizers in gas turbine engines as reported in different publications demonstrates well the excellent stability of silicon carbide. Gas turbine engine components are subjected to extreme conditions of pressure, temperature and vibration. Conventionally machined and macrolaminate atomizers used in gas turbine combustion chambers are designed to survive in such harsh environments where temperatures can reach 1700°C, but they are expensive to manufacture, and both types of atomizers lack the dimensional precision provided by micromachining processes.

As demonstrated, silicon atomizer structures coated with silicon carbide provided a better protection than silicon devices without coating.[116,117] And single crystalline SiC coatings have proven to outperform silicon nitride, silicon dioxide, diamond-like carbon and polycrystalline SiC films regarding wear resistance as demonstrated by Rajan *et al.* in 1998.[117]

Even better results were expected from bulk micromachined silicon carbide devices but the high cost of 6H- and 4H-SiC substrates and the lack of suitable anisotropic and fast etching processes have initially been delaying the development of such devices.

The situation changed when Rajan *et al.* introduced in 1999 a novel process to fabricate bulk micromachined SiC MEMS devices.[118] The approach uses silicon molds fabricated by deep reactive ion etching (DRIE). SiC atomizers were fabricated by high-rate deposition of SiC into such molds. Excess silicon carbide was removed during subsequent polishing steps, and structures were released by dissolving the silicon mold in KOH. The silicon mold and a released silicon carbide fuel atomizer are presented in Figure 9. Injected from an outer chamber, fuel enters the inner spin chamber through four channels and is ejected through the center hole.

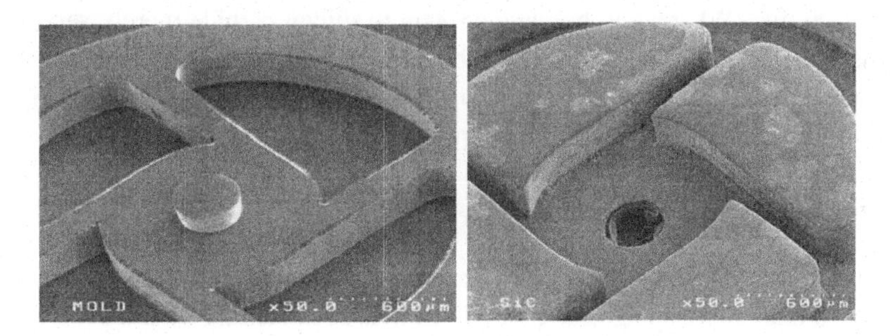

Figure 9. Left: SEM photo of a silicon mold prior to SiC deposition. Right: SEM photo of a SiC atomizer after SiC deposition and dissolving of the silicon mold.[119]

The silicon carbide devices were compared to similar nickel atomizers fabricated using the LIGA process. Performance and erosion tests demonstrated that both types of atomizers are able to perform well at pressures in excess of 2500 kPa, with the SiC devices exhibiting a higher erosion resistance. The results demonstrated a significant improvement over similar silicon devices, which were limited in operation to pressures below 1400 kPa and were comparatively not resistant to erosive wear.

Other coatings for the protection of microfluidic structures based on amorphous silicon carbide were reported by Gallis *et al.* in 2002.[120]

Schmid *et al.* proposed in 2001 the deposition of hydrogenated amorphous silicon carbide (a-SiC:H) as a new passivation layer for the protection of silicon MEMS devices from aggressive media.[121] Etching characteristics and mechanical properties of a-SiC:H films were discussed.

2.7. Chemical sensors

The main efforts on SiC-based chemical sensors can roughly be divided into two categories. In case of field effect-based components e.g. capacitors, Schottky diodes, and field effect transistors (FETs) the adsorption of molecules leads to a change of the electronic properties of SiC films. Whereas in hotplate structures silicon carbide merely serves as a mechanical support and may be used to sense changes of electrical properties of sensitive films e.g. suitable metal oxides deposited on top of the silicon carbide supporting structure.

2.7.1. Field effect gas sensor devices

Field effect sensor devices based on silicon were introduced already more than 30 years ago for example as hydrogen sensors by Lundstroem *et al.* in 1975.[122]

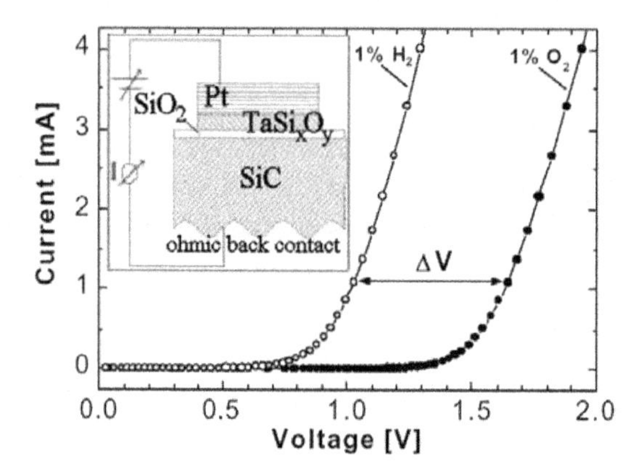

Figure 10. Current-voltage curve of a MISiC diode in two different ambients, 1% H_2 in N_2 and 1% O_2 in N_2, respectively. The sensor signal is the voltage at the metal gate, which is necessary to keep the current constant at 1 mA; the substrate is grounded. The sensor responds to a change in the ambient gas composition with a voltage shift, indicated in the figure as ΔV. The insert shows the structure of the sensor, the insulator SiO_2 is very thin (1-3 nm).[131]

The sensing mechanism is based on hydrogen atoms from reducing hydrocarbon or hydrogen molecules which dissociate at a catalytic metal surface. Hydrogen atoms diffuse through the metal and form an electrically polarized layer at the metal/semiconductor interface. They lower the flat band voltage of capacitors and the barrier height of diodes. Oxygen atoms and e.g. the oxygen of dissociated nitric oxide molecules consume hydrogen and have the opposite effect. Carbon monoxide on the other hand tends to bind oxygen and supports the effect of hydrogen. Therefore, a variety of different gases is detectable with hydrogen sensitive devices.

The advances in SiC material development and electronic devices based on SiC in the middle of the 1990s[123] gave rise to research on gas sensors based on SiC capacitors, diodes and FETs. Targeted applications range nowadays from sensor arrays for food quality control, electronic noses/tongues for aroma analysis to high temperature gas sensors, e.g. for exhaust and flue gas analysis and combustion control.

Advantages of silicon carbide are the chemical inertness, higher operating temperature without device oxidation and degradation of mechanical properties or metal/semiconductor interdiffusion and fast sensor response time due to the accelerated adsorption and desorption of molecules at higher operating temperatures. Different recent overviews of silicon carbide-based gas sensors were presented.[124-128]

In 1997 for example, Lloyd Spetz gave an introduction to high temperature capable catalytic metal-insulator-silicon carbide (MISiC) sensor devices realized as capacitors and Schottky diodes.[129] Sensitivities of sensors to hydrogen, oxygen, propane, ethane and butane were investigated. A maximum operation temperature of 1000°C was obtained for 4H-SiC capacitors and sensors worked routinely for several weeks at 600°C. Time constants for the response to gases were of the order of milliseconds. The layer sequences of sensors were based on the Pt/$TaSi_x$/SiO_2/SiC system. Interdiffusion of metal and insulator layers at temperatures up to T = 600°C were studied.

In general, diodes are preferred over capacitors due to much simpler readout electronics, although capacitors show much better long-term stability due to thick oxide films between metal and semiconductor. Figure 10 depicts the structure of a MISiC diode and its sensitivity towards changes in oxygen and hydrogen concentration of the nitrogen atmosphere as demonstrated by Tobias and Lloyd Spetz *et al.* in 2000.[130]

Based on a similar SiC Schottky diode structure, Baranzahi *et al.* presented in 1998 a combustion monitoring gas sensor operated up to T = 700°C.[132] The sensor was exposed to the exhaust gas of a 4-cylinder petrol engine and the response to excess fuel in one of the cylinders is shown in Figure 11. Due to the fast response within milliseconds, it was possible to distinguish signals from single cylinders. Sensor devices with some selectivity to nitric oxide in synthetic diesel exhaust gases, and devices with high selectivity to carbon monoxide in flue gases were presented as well.

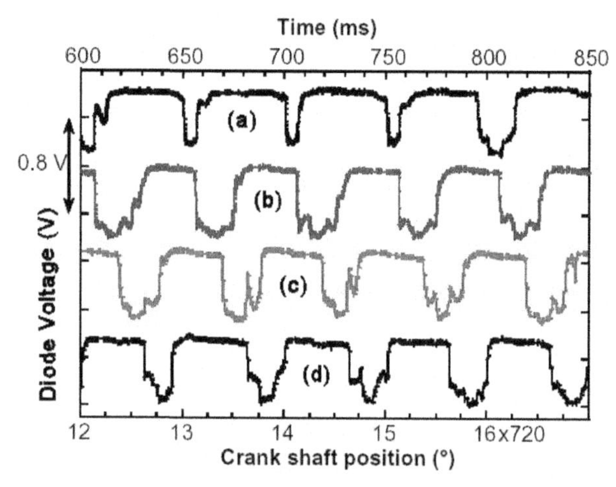

Figure 11. Response of a Schottky diode placed in the manifold of a 4-cylinder gas engine running at 2400 rpm. The upper curve (a) shows the sensor signal for cylinder 1 running fat (excess fuel) and the other three cylinders running close to normal. Next curve (b) cylinder 3 is running fat, next curve (c) cylinder 4 is running fat and bottom curve (d) cylinder 2 is running fat. Sensor temperature ~700°C.[133]

The choice of catalytic metal, structure of the metal, and the operation temperature determines the response pattern to different gases. Svenningstorp *et al.* made use of this fact and generated in 2001 arrays of different MISiC diodes for the analysis of the hydrocarbon content in exhaust gas mixtures.[134] SiC-based chemical sensor arrays were also discussed by Hunter *et al.* in 2002.[135]

The role of hydrogen in these applications is essential and investigations were carried out to study the diffusion of hydrogen in p- and n-type Schottky diodes[136] and the metal-contact enhanced incorporation of deuterium in 4H- and 6H-SiC.[137] Further detailed research was performed on the influence of gate material and operating temperature on Schottky diode properties,[138] the effect of HF and ozone treatment of silicon carbide surface on the performance and stability of diode structures,[139,140] the response of a diode to hydrogen in presence of oxygen,[141] the influence of epitaxial layers fabricated by different methods and with different thickness and doping levels,[142] the influence of catalytic reactivity on the response of MISiC sensors[143] and the role of different gate metal morphologies.[144] The stability of Pt-I-SiC diodes was investigated,[145] the Pt/CeO_2/SiC diode was tested for high sensitivity towards hydrogen,[146] the surface and interface properties of PdCr/SiC Schottky diodes were studied,[147] interfacial and electronic properties of Pd/SiC and Pd/SiO_2/SiC diode systems were reported.[148] The sensitivity of diodes towards CO was researched[149] and hydrogen and hydrocarbon sensors for aeronautical applications based on Pd/SiC Schottky diodes were demonstrated.[150] Schmeisser *et al.* emphasized in 1999 the fact that nanoparticles with an associated dipole moment contribute to the modification of properties of homogeneous interfaces.[151]

FETs combine the advantages of capacitors (better long term stability) and diodes (simple readout electronics). Savage *et al.* gave in 2000 an introduction to gas sensing 4H-SiC FETs.[152] Lloyd Spetz *et al.* gave, in 2000 and 2002 respectively, an overview of MISiCFETs and their applications.[153,154]

Svenningstorp *et al.* demonstrated in 2000 the operation of MISiCFET devices[155] as sensors for NH_3 in diesel exhaust.[156] Such sensors are supposed to support "selective catalytic reduction" (SCR), a method to reduce harmful NO_x by injecting ammonia (NH_3) into the exhaust pipe of a diesel engine.

Wingbrant *et al.* investigated in 2003 the use of MISiCFETs for exhaust and flue gas applications[157] and suggested the use of a MISiCFET as a Lambda-sensor that could be effective even at cold starts when a conventional ZrO_2 sensor cannot properly reduce exhaust emission.[158]

One of the latest developments regarding SiC gas sensors is the development of tunnel Schottky devices using the compound conductor $BaSnO_3$ as a contact as presented by Cerda *et al.* in 2003.[159]

2.7.2. Microhotplates

The basic idea of a microhotplate is to reduce the size and thermal capacitance of a sensing device so that integrated heater elements can control rapidly the operating temperature of the sensor with a minimum level of electrical power.[160] Usually, gas sensing films e.g. metal oxides are deposited on top of sensing electrodes. The output signal refers to changes in capacitance or resistance of the sensing film. The deposition of different films for example through different shadow masks can be used to create gas sensing microhotplate arrays. This way it is possible to mass-produce low-cost, low-power, multi-channel gas detectors e.g. for industrial, environmental and medical purposes.

Silicon carbide can be used to fabricate microhotplates with integrated heater elements and sensing electrodes that do not mechanically degrade or oxidize at high temperatures.

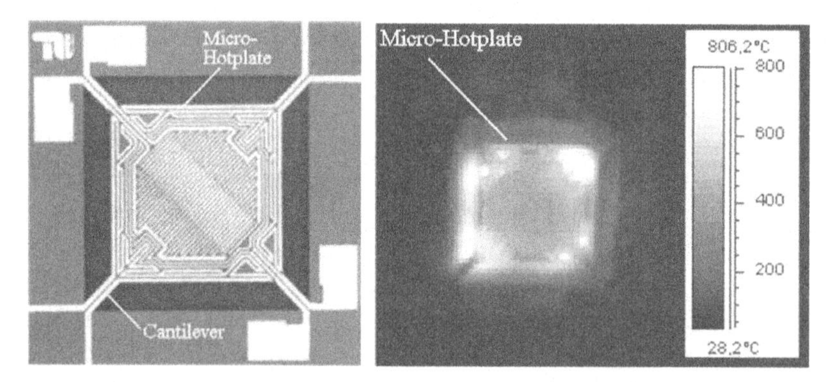

Figure 12. Left: SEM picture of a 600 nm thin SiC microhotplate suspended in air with contacts to the interdigitated capacitor electrodes (top left corner), contacts to the temperature sensing resistor (bottom right corner) and contacts to heater elements (other corners). Right: Thermographic image of the heated microhotplate.[162]

Wiche *et al.* have presented in 2004 a SiC microhotplate with a size of 1x1 mm^2 as shown in Figure 12, left. The silicon substrate underneath the hotplate is removed by KOH etching. Heater elements, temperature sensing resistors and interdigitated capacitor electrodes are made from a sputtered platinum thin film.[161] The thermographic image (Figure 12, right) indicates a uniform hotplate temperature within 22 K at an operating temperature of T = 800°C.

Solzbacher *et al.* have reported already in 1999-2001 on the development of a similar SiC-based microhotplate with a size of 80 x 80 μm^2.[163-168] The design enables the operation at low voltages of 1-2 V. The sensor consists of a 1 μm thick polycrystalline 3C-SiC membrane and a 200 nm thick HfB_2 thin film heater. The heater is insulated from the sensing interdigitated capacitor through a 400nm thick PECVD-SiO_2 layer. HfB_2 limits the maximum operating temperature to T = 650°C. The response of a sensor with a sputtered 120 nm thin In_2O_3 film to an NO_2 concentration of 5 ppm at T = 250°C is shown in Figure 13. The sensor survived more than 20,000 one-minute-long work cycles.

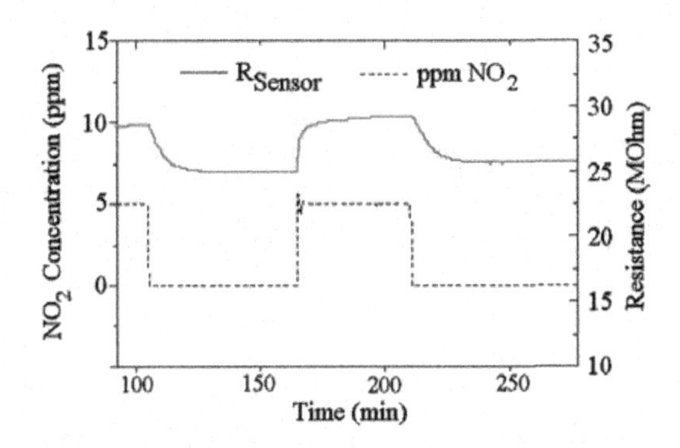

Figure 13. Response of a microhotplate gas sensor to 5 ppm NO_2 at 250°C (20mW heater power). A 120 nm thin In_2O_3 gas-sensitive layer was used.[165]

An ethanol gas sensor was presented by Ho *et al.* in 1998.[169] Silicon carbide was used to fabricate a bridge serving as a mechanical support for the sensing SnO_2 film and as a heater element, respectively. Operation at T = 300°C enabled ethanol detection within 10 seconds.

2.7.3. Other chemical sensors

Relative humidity sensors based on porous silicon carbide films were presented by Connolly *et al.* in 2002.[170] Silicon carbide provided a good etch stop during KOH etching and could enable the use of the sensors in harsh environments. A typical increase in capacitance of the porous layer by a factor of 2-3 was demonstrated as a response to an increase in relative humidity from 0% to 95%. Typically, the response to humidity changes occurred within minutes.

2.8. Optical devices

Light emitting diodes have initially been the first silicon carbide-based products presented by Cree Inc. But since silicon carbide is an indirect bandgap semiconductor, the quantum efficiency and hence the achievable optical output power are relatively low. SiC-based LEDs are nowadays replaced by devices based on GaN and InGaN thin films that are heteroepitaxially deposited either on silicon carbide or sapphire substrates.[171] Silicon carbide photodiode structures are of interest since they extent the detectable spectrum of silicon photodetectors into the ultraviolet region. Other optical devices have been discussed, e.g. SiC-based solar cells and photoconductors. Hydrogenated amorphous silicon carbide films (a-SiC:H) were investigated already in the 1980s, e.g. for their use in solar cells, LEDs and photodiodes. Recent work on photodiodes, image sensors and photoconductors based on a-SiC:H films are briefly discussed below.

2.8.1. Light emitting diodes

As a light emitter, 6H-SiC junctions can be tailored to emit light across the visible spectrum. Already in 1986 Dmitriev *et al.* demonstrated a red-green-blue LED display made from a 1 x 1 cm^2 single crystal wafer of 6H-SiC. Ion implantation and epitaxial techniques were employed to achieve the three LED colors.[172]

Edmond *et al.* gave in 1997 an overview of LEDs fabricated by Cree which were the only commercially available wide bandgap LEDs at that point.[173] One proposed light generation mechanism for these LEDs is phonon assisted donor-acceptor pair recombination between nitrogen donors and aluminum acceptors. This doping scheme resulted in blue LEDs emitting light with a peak wavelength of 470 nm with a spectral halfwidth of ~70 nm. The optical power output was typically between 25μW and 35μW at a forward current of ~20mA at 3.2V. This represents an external quantum efficiency of ~0.05 to 0.07%. Green LEDs have been demonstrated which emit with a peak wavelength of 530 nm. As opposed to the epitaxial junction in blue LEDs, the green devices used ion-implanted junctions. The green luminescence is attributed to defects created close to the junction during implantation.

2.8.2. Photodetectors

In addition to short wavelength light emission as discussed above, the wide bandgap allows for inherently low dark currents and high quantum efficiencies for ultraviolet photodiode detectors, even at high temperatures.

As pointed out by Edmond *et al.*, high temperature SiC UV photodiodes may have a significant impact in many application areas. Improvement in combustion control is anticipated with the ability to sense flames in aircraft engines, building boiler systems and industrial processes. Air quality monitoring and UV dosimetry for industrial processes are other important application areas. A small, low-power, temperature insensitive UV detector would enable the construction of rugged, self-contained UV detector systems for use in remote-sensing applications. The availability of an array detector, sensitive to the near and middle UV, could be important to imaging applications, including atmospheric UV remote sensing, combustion control and potentially missile plume detection and tracking.[173]

Photodiodes as fabricated by Cree typically exhibited a quantum efficiency of 80 to 100% and peak response of 250 to 280 nm.[173] These characteristics were maintained to at least 350°C (Figure 14, left). The dark current density at -1.0V and 473K was $\sim 10^{-11}$ A/cm^2. This corresponds to an extrapolated room temperature dark current density of $\sim 2 \times 10^{-17}$ A/cm^2 at -1.0V.

Brown *et al.* have presented the use of a silicon carbide photodiode as a sensor for combustion engine control in 1996.[175] The UV emission spectrum of hydrocarbon flames is well matched to the optical properties of SiC.

Another important characteristic of a photodiode is the ability to generate power when illuminated; it can be operated as a solar cell. Figure 14, right shows the I-V characteristics of a 6H-SiC photodiode. The shaded area indicates the maximum power rectangle. The maximum power P_m occurs at $I_m = 0.35$mA and $V_m = 1.35$V, or $P_m = 0.47$mW.

Hubbard and Raffaelle *et al.* investigated the impact of crystal defects on the performance of SiC solar cells and suggested the use of SiC-based solar cells for high-temperature, high-light intensity and high-radiation missions, such as experienced by solar probes.[176,177]

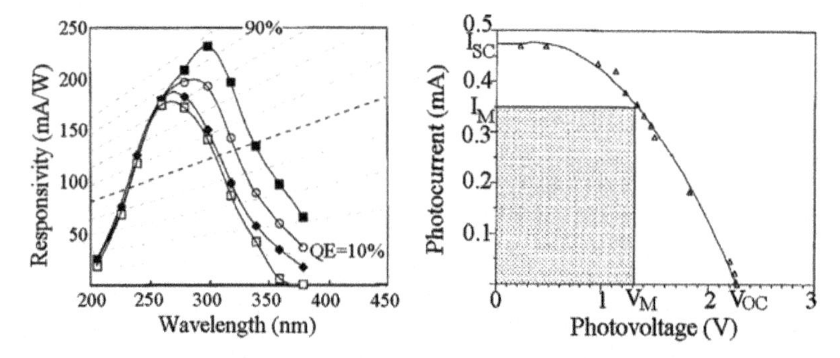

Figure 14. A photodiode can either be used to detect light or to produce power during illumination. Left: the effect of measurement temperature on the responsivity and quantum efficiency of a SiC UV detector. The temperatures tested were □ 223K, ♦ 300K, o 498K, and ■ 623K. Right: relationship between the generated photocurrent vs. photovoltage for a 3 x 3 mm^2 6H-SiC photodiode. The incident light energy was 7×10^{-3} W/cm^2 with wavelengths in the range of 200 to 350 nm. The shaded area represents the maximum power rectangle of the device.[174]

A different type of photodetector is based on hydrogenated amorphous material. Color sensors using hydrogenated amorphous silicon (a-Si:H) are of interest because of their high responsivity in the visible spectrum, low-cost fabrication, and homogeneous deposition over large areas by plasma-enhanced chemical vapor deposition (PECVD).[178]

Cabrita *et al.* presented similar sensors fabricated as Schottky photodiodes based on the material system glass/ITO/a-Si_x:C_{1-x}:H/Al in 2001.[179]

Glass serves as the transparent substrate, indium tin oxide (ITO) is a transparent and conductive film that provides the back contact, while aluminum serves as the front contact to the hydrogenated amorphous silicon carbide (a-SiC:H) structure. Responsitivities of these diodes of the order of several 10 mA/W were recorded. Switching the diode bias voltage from −1V to +1V resulted in a shift of the peak of the spectral response from 630 nm to 540 nm (Figure 15).

Color selective position sensitive detector (PSD) devices[181] and laser scanning photodiode (LSP) image sensors[182] based on a–SiC:H films were realized by Cabrita *et al.* and Vieira *et al.* in 2001, respectively. An LSP image sensor consists of a single, large area p–i–n sensing element and an optomechanical acquisition system. The image to be acquired is optically mapped on the photosensitive surface and a low-power chopped laser scans the sensor in raster mode. The detection of generated carriers

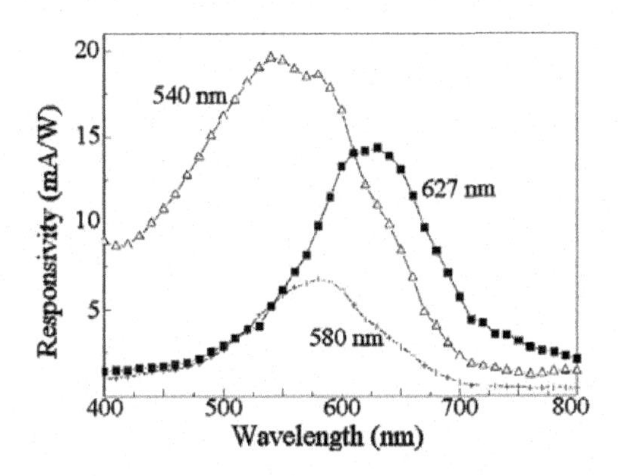

Figure 15. Spectral response of ITO/a-Si_x:C_{1-x}:H/Al photodiodes under different bias voltages (■ = -1V, + = 0V, Δ = +1V).[180]

is achieved by measuring the AC-component of the short circuit current that depends on the intensities of both the image and scanner beam.

Vlaskin *et al.* have discussed the efficiency of photoconductivity of amorphous silicon carbide films in 2003.[183] An effective number of electrons (N_{eff}) was defined as the number of electrons taking part in the conductivity per 100 incident light photons. A maximum value of $N_{eff} = 10\text{-}12$ was measured at $\lambda = 540$ nm. Amorphous silicon carbide has also been used very early for the fabrication of photoluminescence displays[184] and solar cells.[185]

In an earlier work in 1995, Cho *et al.* investigated the photoconductive and photovoltaic response of high dark-resistivity 6H-SiC devices. A switching efficiency of pn-junction-based photoconductive switches of more than 80% was achieved.[186] Similar structures were used by Saddow *et al.* to create sub-nanosecond photovoltaic electrical pulses by illumination of a 6H–SiC photoconductive switch device with a nitrogen laser. The use in integrating optically activated test circuit structures with fast electronic devices was indicated as a potential application.[187]

2.8.3. Other optical devices

Cheng *et al.* presented in 2003 the fabrication and characterization of Fabry–Perot interferometers based on polycrystalline SiC films grown on single crystalline sapphire substrates.[26] SiC film thickness and surface roughness values were in the range of 0.5 – 2 μm and 1 – 8.5 nm, respectively. These sensors were designed for fiber-optic temperature measurements in harsh environments. Using a white, broadband light source, a temperature accuracy of 3.5°C was obtained over the temperature range of 22°C to 540°C.

2.9. Radiation detectors

SiC-based nuclear radiation detectors were investigated already in the 1960s but other materials, e.g. Ge, Si, CdTe, showed much better dynamics of necessary material improvement compared to SiC.

A large number of research results on SiC-based radiation detectors has been published. As an introduction, recent achievements by Strokan *et al.* are briefly discussed.

Strokan *et al.* pointed out in 2003[188] that recent success in control over SiC material properties has made it possible to obtain a combination of material characteristics necessary for successful detector operation: low impurity concentration (extended electric field region in a diode structure typically used in detectors), bipolar conduction (no accumulation of a space charge distorting the field), long drift length of charge carriers (carrier transport with efficiency close to unity), wide bandgap ensuring weak thermal generation of carriers (low noise), and possibility of creating high-voltage diode structures. Since tracks of nuclear particles (quanta) occupy only a small part of the detector volume, high local uniformity of carrier transport conditions throughout the detector volume is necessary.

The presently achieved lower limit of impurity concentration in n-type SiC is 5×10^{14} cm^{-3} – 1×10^{15} cm^{-3}. This results in field regions with a width of 30 μm at a voltage of 500V. Carrier lifetimes of the order of hundreds of nanoseconds for holes combined with high values of saturated drift velocity ensure a nearly 100% efficiency of charge transfer. Particularly attractive are the radiation hardness, chemical stability of SiC, and also the possibility of device operation at high temperatures.

Strokan *et al.* presented triode device structures based on thick and thin SiC epitaxial films as detectors of alpha particles and weak ionization radiation (x-ray and UV quanta), respectively.[188] Alpha particle spectrometry was demonstrated. The triode structure resulted in an amplified detection of x-ray and optical (UV) quanta by a factor of several tens. A superlinear rise in the resulting signal was observed with increasing voltage.[188]

2.10. RF MEMS

Radio Frequency (RF) MEMS devices have a broad range of potential applications in military and commercial wireless communication, navigation and sensor systems. An introduction to this emerging field was given for example by Richards and De Los Santos in 2001.[189,190]

RF MEMS devices include switches, inductors and variable capacitors, high Q filters, phase shifters, antennas with reconfigurable radiation characteristics and transmission lines.

Melzak pointed out in 2003 that silicon carbide is an excellent candidate for its use in next generation RF MEMS devices.[191] The material properties of silicon carbide are supposed to potential improvements in operating frequency, power handling capability and reliability of such devices relative to their silicon counterparts.

A first step towards RF MEMS based on silicon carbide was taken by Huang *et al.* in 2002.[192] They presented nanomechanical resonators with fundamental mode resonant frequencies in the ultra high frequency (UHF) band. The resonator structures were fabricated from single crystalline 3C–SiC thin films, epitaxially grown on Si substrates using a combination of electron beam lithography and micromachining techniques. The beams were roughly 1.25 μm long, 180 nm wide and 75 nm thick. Resonant frequencies of doubly clamped resonator pairs were measured by magnetomotive transduction[193] combined with a balanced bridge readout circuit. Resonant frequencies as high as 632 MHz were recorded.

2.11. SiC NEMS

According to the Network for Computational Nanotechnology (NCN)[194] "Nanoelectromechanical Systems (NEMS) are nanoscale sensors, actuators, devices, and systems with critical dimensions on the order of a few nanometers (typically less than 100 nm). NEMS could be an enabling technology with great potential for breakthrough developments in science and engineering. NEMS promises fundamental frequencies in the 1-100 GHz range, mechanical quality factors in the range of 10^3 to 10^5, active masses in the femtogram range, force sensitivities at the attonewton level, mass sensitivity at the level of an attogram, short response times, and power consumption in attowatts.

These attributes of NEMS could enable sensing of individual cells, proteins, and DNA; design of low power switches; nanomechanical resonators for ultra-sensitive detection of adsorbed mass; radio frequency devices for computing (specifically nonvolatile RAM); nanotweezers; ultra high data storage; and several other devices. Achieving the promise of NEMS involves more than simply scaling down MEMS; new physical phenomena associated with surfaces, interfaces, and atomic scales must be understood." Silicon carbide will likely contribute with its unique material properties to the success of NEMS devices.

Work on doubly clamped beams has been discussed in the RF MEMS sub-chapter above and can be considered as one of the first steps into the world of SiC NEMS. Yang *et al.* reported on a similar work already in 2001.[195] SEM pictures of a families of resonant beams are shown in Figure 16.

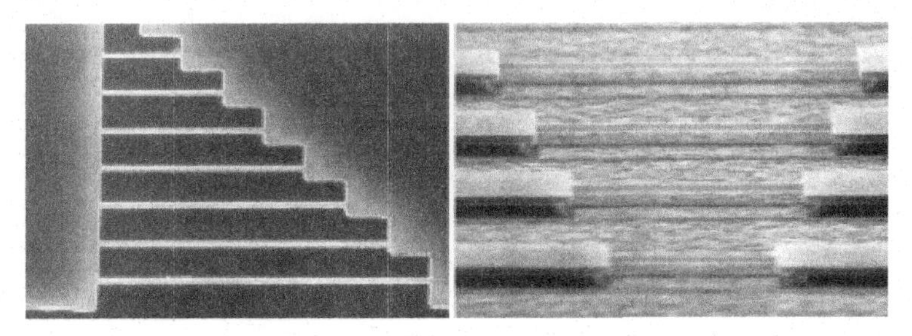

Figure 16. Doubly clamped SiC beams patterned from a 259 nm thick epilayer. Left: top view of a family of 150 nm wide beams, having lengths from 2 to 8 μm. Right: side view of a family of 600 nm wide beams, with lengths ranging from 8 to 17 μm.[196]

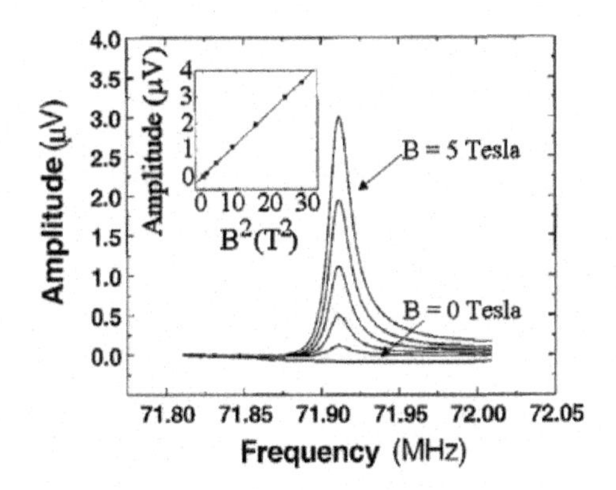

Figure 17. Representative data of a SiC doubly clamped beam resonating at 71.91 MHz, with a quality factor Q = 4000 (at T = 20 K). The family of resonance curves is taken at various magnetic fields; the inset shows the characteristic B^2 dependence expected from magnetomotive detection. For clarity of presentation the data is normalized to response at zero magnetic field, with the electrode's dc magnetoresistance shift subtracted from the data. This provides an approximate means for separating the electromechanical response from that of the passive measurement circuitry.[196]

The fabrication process involved electron beam lithography followed by dry anisotropic and selective electron cyclotron resonance plasma etching steps. Electromechanical characteristics were measured using the magnetomotive detection technique[193] from 4.2 K to 295 K in a superconducting solenoid within a variable temperature cryostat. The resonance curves of beams with a length of 8 μm, a width of 600 nm and a thickness of 259 nm at a resonant frequency of 71.91 MHz are shown in Figure 17.

3. Conclusions and Outlook

Good market prospects and successful development and marketing of the first SiC-based electronic and optoelectronic products have initiated and accelerated the development of SiC material systems and SiC fabrication technologies and have also given rise to the realization of SiC MEMS prototypes.

6H and 4H-SiC wafers with a diameter of up to three inches and per request with epitaxial layers are offered today as commercial products by a variety of companies. A large number of universities and research institutions are able to produce 3C-SiC-based material systems with a satisfying quality. The company FLX Micro introduced the MUSIC® process as a foundry service in 2002. Hence, a powerful 3C-SiC multilayer technology with reproducible SiC thin film quality at an affordable price is now commercially available as well.

More fabrication technologies have emerged and enable the flexible realization of surface and bulk micromachined SiC MEMS devices. SiC-to-SiC wafer bonding processes were demonstrated. Fast wet and dry etching processes and even deep reactive ion etching (DRIE) processes were recently presented. Polishing processes for 3C-SiC films and procedures for the transfer of films to other substrates by wafer bonding were developed. 3C-SiC molding processes for surface and bulk micromachining represent important technologies. Reliable metallization systems for operating temperatures higher than 600°C and high temperature die attach schemes and wire bonding processes are available.

But despite these technological achievements, only electronic and optoelectronic devices are currently offered as commercial products. The blue LED was introduced as the first SiC product in 1989 by Cree Inc. followed by the commercialization of GaN/SiC-based blue, green and near-UV LEDs, SiC MESFET RF devices, Schottky diodes and rectifiers. Further near-term SiC-based product targets include white LEDs, blue lasers for the next-generation DVD market and optical storage and the next generation of devices for power switching and wireless communication.

Prospects for SiC-based electronic and optoelectronic products remain very promising. According to market research firm Strategies Unlimited, the market for high brightness, GaN-based LEDs, for example, is expected to grow at a compound annual growth rate of 25% to $4 billion by 2007. Electronic and optoelectronic SiC products will therefore keep pushing the further development of SiC substrates and fabrication technologies. Decreasing substrate prices, better availability of powerful fabrication processes and increasing research activities will help SiC MEMS devices gain market maturity as well.

Research on high temperature, power and RF electronics, for example, has already accelerated the development of gas sensors based on field effect devices. Numerous studies on field effect capacitors, diodes and transistors for gas sensing purposes have been published. The development of suitable metallization systems has led to the demonstration of gas sensors at operating temperatures of more than 600°C over weeks. Promising fields of applications, e.g. control of gasoline engines and combustion control in general, were addressed. Provided that long term stability and reliability of metallization and packaging solutions can be ensured, the first SiC MEMS product in the near future could well be a gas sensor.

Pressure sensors are main products within the silicon MEMS field. Prototypes based on SiCOI systems were successfully operated up to 400°C. Bulk micromachining of 6H-SiC was applied to fabricate pressure sensors that were operated up to 600°C with impressive signal stability. But at least up to 350-400°C SiC-based pressure sensors will be in competition to SOI-based sensors. SOI represents the more mature and cost effective material system. Unless bulk SiC material drops significantly in price, or the SiCOI material system increases significantly in quality, SiC might only be interesting for high temperature small niche markets above 400-500°C that cannot be covered by SOI at all.

A similar situation is given for accelerometers. DRIE of a 6H-SiC substrate was successfully applied to fabricate a first prototype. But again a suitable, large enough market for accelerometers must be identified that cannot be covered by sensors based on SOI and other material systems.

RF MEMS is an emerging field. According to an estimate by Wicht Technology Consulting the total market for RF MEMS will exceed US $1 billion in 2007.[197] SiC resonators with very high resonant frequencies above 600 MHz have already been presented and a powerful technology as offered by the MUSIC® process might help the SiC MEMS community to enter the race for market shares.

Other presented promising SiC MEMS prototypes include wear resistant fuel atomizers, micromotors, microhotplates, temperature sensors and radiation detection devices. This list well indicates further market potential of SiC MEMS.

Overall SiC MEMS is a very exciting and promising field. It capitalizes on the success of SiC-based electronic and optoelectronic products. The pace of the recent development of SiC MEMS technologies and devices is encouraging. The price for SiC material has to decrease further. Improvements regarding material quality and SiC MEMS fabrication and packaging technologies are necessary to ensure reliability and competitiveness of SiC MEMS products. From a product point of view the field of SiC MEMS is still in its infancy but it certainly has the potential for a successful future.

4. Acknowledgment

The author is thankful to Dr. Rebecca Cheung (University of Edinburgh) for her kind invitation to contribute to this book, Prof. Ernst Obermeier (Berlin University of Technology) for his support during the author's time as a graduate student when he was working on SOI and SiC-based pressure sensors and Prof. Olav Solgaard (Stanford University) for his support during the production of this chapter's manuscript.

REFERENCES

1. Bardeen, J., Brattain, W.H., *Phys. Rev.,* **74,** (1948), pp.230-231.
2. Petersen, K.E., *Proc. of the IEEE*, **70,** (1982), pp.420-57.
3. Petersen, K.E., *Conf. Proc. Transducers'95*, Stockholm, Sweden (1995), pp.894-7.
4. Krötz, G., Legner, W., Wagner, C., Moller, H., Sonntag, H. and Muller, G., *Conf. Proc. Transducers'95*, Stockholm, Sweden (1995), pp.186-9.
5. Krötz, G., Möller, H., Eickhoff, M., Zappe, S., Ziermann, R., Obermeier, E. and Stoemenos J., *Mat. Science and Eng.,* **B61-62,** (1999), pp.516-521.
6. Mehregany, M. and Zorman, C.A., *Silicon Carbide 2002 – Mat. Proc and Dev*., ed. Agarwal A. *et al.*, Vol. 640, (MRS, Warrendale, PA, USA, 2002), *H4.3.1-6.*
7. Kroetz, G.H., Eickhoff, M.H. and Moeller, H., *Sensors and Actuators A (Physical),* **A74**, (1999), pp.182-9.
8. Liu, S., Tang, Y. and Li, Si-yuan, *Micronanoelectronic Technology,* **39**, (2002), pp.26-30.
9. Mehregany, M., Zorman, C.A., Roy, S., Fleischman, A.J., Wu, C.-H. and Rajan, N., "Silicon carbide for microelectromechanical systems", *International Materials Reviews*, **45**, (2000), pp.85-108.
10. Mehregany, M., Zorman, C.A., Rajan, N. and Wu, C.H., *Proc. of the IEEE*, **86,** (1998), pp.1594-609.
11. Mehregany, M. and Zorman, C.A., *Thin Solid Films,* **355-356,** (1999), pp.518-24.
12. Zorman, C.A. and Mehregany, M., *Conf. Proc. IEEE SENSORS 2002*, Orlando, FL, USA (2002), pp.1109-14.
13. Bauer, K., Kroetz, G., Schalk, J. and Mueller, G., *Proc. of the SPIE,* **4698**, (2002), pp.497-508.
14. Carter, C.H. Jr., Edmond, J.A. and Palmour, J.W., *Conf. Proc. SAMPE 7th International Electronic Materials and Processes Conf.*, Parsippany, NJ, USA, (1994), pp.8-17.
15. Yoshioka, M., *J. Appl. Phys.,* **76,** (1994), pp.7790-7796.
16. Markus, K.W., Koester, D.A., Cowen, A., Mahadevan, R., Dhuler, V.R., Roberson, D. and Smith, L., *Proc. of the SPIE,* **2639,** (1995), pp.54-63.

17. Berzelius, J.J., *Ann. Phys.*, Lpz., **1**, (1824), pp.169.
18. Acheson, E.G., "On Carborundum", *Chem. News,* **68**, (1893), p.179.
19. Lely, J.A., Ber. dtsch. Keram. Gesellschaft, **32**, (1955), p.229.
20. Tairov, Y.M. and Tsvetkov, V.F., *J. Cryst. Growth*, **43**, (1978), p.209.
21. Davis, R.F., Carter, C.H. and Hunter, C.E., US Patent No. 4866005: September 12, (1989).
22. Nishino, S., Hazuki, Y., Matsunami, H. and Tanaka, T., *J. of the Electrochem. Society*, **127**, (1980), pp.2674-80.
23. Matsunami, H., Nishino, S. and Ono, H., *IEEE Trans. Electron Devices*, ED-28, (1981), p.1235.
24. Kong, H., Kim H.J., Edmond, J.A., Palmour, J.W., Ryu, J., Carter, Jr., C.H., Glass, J.T. and Davis R.F., *Conf.: Novel Refractory Semicon. Symp.,* Anaheim, CA, USA (1987), pp.223-45.
25. Camassel, J., *J. Vac. Sci. Technol.,* **B16**, (1998), pp.1648-54.
26. Cheng, L., Steckl, A.J. and Scofield, J., *IEEE Trans. Electron Devices*, **50**, (2003) pp.2159-64.
27. Luo, M.C., Li, J.M., Wang, Q.M., Sun, G.S., Wang, L., Li, G.R., Zeng, Y.P. and Lin, L.Y., *J. of Crystal Growth,* **249,** (2003), pp.1-8.
28. Cheng, L., Pan, M., Scofield, J. and Steckl, A.J., *J. of Electronic Materials*, **31,** (2002), pp.361-5.
29. Serre, C., Perez-Rodriguez, A., Morante, J.R., Esteve, J., Acero, M.-C., Kogler, R. and Skorupa, W., *J. of Micromechanics and Microengineering,* **10**, (2000), pp.152-6.
30. Serre, C., Perez-Rodriguez, A., Romano-Rodriguez, A., Morante, J.R., Esteve, J., Acero, M.C., Kogler, R. and Skorupa, W., *Conf. Progress in SOI Structures and Devices Operating at Extreme Conditions, Proc. of the NATO Adv. Res. Workshop,* Kyiv, Ukraine (2000).
31. Vinod, K.N., Zorman, C.A. and Mehregany, M., *Conf Proc. Transd. '97,* Chicago, IL, USA, (1997), pp.653-6.
32. Zorman, C.A., Vinod, K.N., Yasseen, A. and Mehregany, M., *Materials Science Forum,* **264-268**, (1998), pp.223-6.
33. Di Cioccio, L., Le Tiec, Y., Letertre, F., Jaussaud, C. and Bruel, M., *Electronics Letters*, **32,** (1996), pp.1144-5.
34. Di Cioccio, L., Le Tiec, Y., Jaussaud, C., Hugonnard-Bruyere, E. and Bruel, M., *Materials Science Forum*, **264-268**, (1998), pp.765-70.
35. Song, X., Rajgopal, S., Melzak, J.M., Zorman, C.A. and Mehregany, M., *Materials Science Forum*, **389-393,** (2002), pp.755-8.
36. Wu, C.H., Chung, J., Hong, M.H., Zorman, C.A., Pirouz, P. and Mehregany, M., *Mat. Science Forum*, **353-356,** (2001), pp.171-4.
37. Eickhoff, M., Zappe, S., Nielsen, A., Krotz, G., Obermeier, E., Vouroutzis, N. and Stoemenos, J., *Materials Science Forum,* **353-356,** (2001), pp.175-8.

38. Reichert, W., Stefan, D., Obermeier, E. and Wondrak, W., *Mat. Sci. Eng.,* **B46,** (1997), pp.190-4.
39. Jiang, L., Cheung, R., Hassan, M., Harris, A.J., Burdess, J.S., Zorman, C.A. and Mehregany, M., *J. Vac. Sci. Technol.,* **B21,** (2003), p.2998.
40. Pearton, S., *Process technology for silicon carbide devices,* ed. Zetterling C.-M., (London, INSPEC, 2002), pp.85-92.
41. Yih, P.H., Saxena, V. and Steckl, A.J., *Physica Status Solidi, B,* **202,** (1997), pp.605-642.
42. Beheim, G. and Salupo, C.S., *Conf.: Wide-Bandgap Electronic Devices. Symp.,* San Francisco, CA, USA (2000), T8.9.1-6.
43. Shor, J.S. and Kurtz, A.D., *J. of the Electrochemical Society,* **141,** (1994), pp.778-81.
44. Chang, W.-H., Schellin, B., Spellmeyer, R., Obermeier, E. and Huang, Y.-C., *Conf. Proc. Eurosensors XVI,* Prague, Czech Republic, (2002), p.246.
45. Shor, J.S., Zhang, X.G. and Osgood, R.M., *J. of the Electrochem. Soc.*, **139**, (1992), pp.1213-16.
46. Shor, J.S., Osgood, R.M. and Kurtz, A.D., *Applied Physics Letters*, **60,** (1992), pp.1001-3.
47. Shor, J.S., Kurtz, A.D., Grimberg, I., Weiss, B.Z. and Osgood, R.M., *J. of Applied Physics*, **81,** (1997), pp.1546-51.
48. Dong, Y., Zorman, C. and Molian, P., *J. of Microm. and Microeng.*, **13,** (2003), pp.680-5.
49. Bischoff, L., Teichert, J. and Heera, V., *Applied Surface Science*, **184,** (2001), pp.372-376.
50. Yasseen, A.A., Zorman, C.A. and Mehregany, M., *J. of Microelectromechanical Systems*, **8,** (1999), pp.237-42.
51. Rajan, N., Zorman, C.A. and Mehregany, M., *Materials Science Forum*, **338-342,** (2000), pp.1145-8.
52. Fu, X., Zorman, C.A. and Mehregany, M., *J. of the Electrochemical Society*, **149,** (2002), G643-7.
53. Bruel, M., Aspar, B. and Auberton-Hervé, A.J., *Jpn. J. Appl. Phys.*, **36,** (1997), pp.1636-41.
54. Yushin, G.N., Kvit, A.V., Collazo, R. and Sitar, Z., *Silicon Carbide 2002 – Mat. Proc and Dev.*, ed. Saddow S. E. *et al.* (MRS, Warrendale, PA, USA 2002), pp.91-95.
55. Porter, L.M. and Davis R.F., *Mat. Sci. Eng.,* **B34,** (1995), pp.83-105.
56. Jang, T., Odekirk, B., Madsen, L. D. and Porter, L. M., *J. Appl. Phys.,* **90,** (2001), p.4555.
57. Chen, L.Y., Hunter, G.W. and Neudeck, P.G., *Conf.: Wide-Bandgap El. Devices. Symp.,* San Francisco, CA, USA, (2000), T8.10.1-6.

58. Burla, R. K., Roy, S., Haria, V. M., Zorman, C. A. and Mehregany M., Proc. Conf. Design, Characterization, and Packaging for MEMS and Microlectronics, SPIE Symp. on Microelectronics and MEMS, **3893,** (1999), pp.324–333.
59. Senturia, S., *Microsystem Design*, (Kluwer, 2001).
60. Madou, M., Fundamentals of Microfabrication, (CRC, 2002).
61. Kovacs, G., Micromachined Transducer's Sourcebook, (McGraw-Hill, 1998).
62. Maluf, N., An Introduction to Microelectromechanical Systems Engineering, (Artech House, 2000).
63. Rebeiz, G.M., *RF MEMS – Theory, Design, Technology*, (John Wiley and Sons, 2002).
64. Rai-Choudhury, P., (Ed.) *MEMS and MOEMS, Technology and Applications*, (SPIE Press, 2000).
65. Koch, M., Evans, A. and Brunnschweiler A., *Microfluidic Technology and Applications*, (Research Studies Press, 2000).
66. Mehregany, M., Tong, L., Matus, L. G. and Larkin D. J., *IEEE Trans. Electron Dev.*, **44,** (1997), pp.74-79.
67. Serre, C., Perez-Rodriguez, A., Romano-Rodriguez, A., Morante, J.R., Esteve, J. and Cruz Acero, M., *J. of Micromechanics and Microengineering*, **9,** (1999), pp.190-3.
68. Figures reprinted from ref. 67 with permission from *Institute of Physics Publishing*
69. Smith, C.S., *Phys. Rev.*, **94,** (1954), pp.42-49.
70. Shor, J.S., Goldstein, D. and Kurtz, A.D., *IEEE Trans. on Electron Devices*, **40,** (1993), pp.1093-9.
71. Keyes, R.W., *Solid State Physics*, **11** (New York, Academic, 1960), p.1493.
72. Reichert, W., González Sirgo, M., von Berg, J., Obermeier, E., *Trans. of the 3rd Int. HiTEC Conf.*, Vol. 2, Albuquerque, NM, USA, June 9-14, (1996) pp. 137-42.
73. Shor, J.S., Bemis, L. and Kurtz A.D., *IEEE Trans. on Electron Devices*, **41,** (1994), pp.661-5.
74. Okojie, R.S., Ned, A.A., Kurtz, A.D. and Carr W.N., *IEEE Trans. on Electron Device,* **45,** (1998), pp.785-90.
75. Jansen, E., Ziermann, R., Obermeier, E., Kroetz, G., Wagner, Ch., *Material Science Forum* **264-8,** (1998), pp.631-4.
76. Irace, A. and Sarro, P. M., *Conf. Eurosensors XIII*, The Hague, Netherlands (1999), 25P2.
77. King J. A. (Ed.), *Materials Handbook for Hybrid Electronics*, (Artech House, Boston/London, 1988), p.586.
78. Ziermann, R., von Berg, J., Obermeier, E., Wischmeyer, F., Niemann, E., Möller, H., Eickhoff, M. and Krötz, G., *Material Science and Engineering*, **B61-62,** (1999), pp.576-578.
79. von Berg, J., Ziermann, R., Reichert, W., Obermeier, E., Eickhoff, M., Krötz, G., Thoma, U., Cavalloni, C. and Nendza J.P., *Proc. of the HITEC '98*, Albuquerque, NM, USA, (1998), pp.245-249.

80. von Berg, J., Ziermann, R., Reichert, W., Obermeier, E., Eickhoff, M., Krötz, G., Thoma, U., Boltshauser, Th., Cavalloni, C. and Nendza J.P., *Materials Science Forum,* **264-268,** (1998), pp.1101-4.
81. Figure(s) reprinted from ref. 80 with permission from *Trans Tech Publications.*
82. Ziermann, R., von Berg, J., Reichert, W. and Obermeier, E., *Proc. Transducers 97*, Chicago, Illinois, USA, (1997), pp.1411-14.
83. von Berg, J., Ziermann, R., Reichert, W., Obermeier, E., Eickhoff, M., Krötz, G., Thoma, U., Boltshauser, Th., Cavalloni, C. and Nendza J.P., *Proc. ICSCIII - N'97*, Stockholm, Sweden (1997), p.520.
84. Ziermann, R., von Berg, J., Obermeier, E., Wischmeyer, F., Niemann, E., Moller, H., Eickhoff, M. and Krotz, G., *Materials Science & Engineering,* **B61-B62,** (1999), pp.576-8.
85. Zappe, S., Obermeier, E., Eickhoff, M., Möller, H., Krötz, G., Bonnotte, E., Barriol, Y., Decorps, J.L., Rougeot, C., Lefort, O. and Stoemenos, J., *Conf. HITEC 2000,* Albuquerque, NM, USA, (2000).
86. Zappe, S. (Dissertation, ISBN 3-18-335809-3, VDI Verlag GmbH, Duesseldorf, Germany, 2003).
87. Zappe, S., Obermeier, E., Möller, H., Krötz, G., Bonnotte, E., Barriol, Y., Decorps, J.L., Rougeot, C., Lefort, O. and Menozzi, G., *Conf. Transducers '99*, Sendai, Japan (1999), pp.346-9.
88. Zappe, S., Eickhoff, M. and Stoemenos, J., *Conf.: Microelectronics, Microsystems and Nanotechnology,* Athens, Greece, (2000), pp.227-33.
89. Zappe, S., Franklin, J., Obermeier, E., Eickhoff, M., Moller, H., Krotz, G., Rougeot, C., Lefort, O. and Stoemenos, J., *Materials Science Forum*, **353-356,** (2001), pp.753-6.
90. Möller, H., Zappe, S., Papaioannou, V., Krötz, G., Skorupa, W., Obermeier, E. and Stoemenos, J., *Proc. ECS Meeting,* Seattle, USA, **MA 99-1,** (1999), p.398.
91. Zappe, S., Moller, H., Krotz, G., Eickhoff, M., Skorupa, W., Obermeier, E. and Stoemenos, J., *Materials Science Forum*, **338-342,** (2000), pp.529-32.
92. Zappe, S., Obermeier, E., Stoemenos, J., Möller, H., Krötz, G., Wirth, H. and Skorupa, W., *Mat. Science and Eng.,* **B61-62,** (1999), pp.522-5.
93. Eickhoff, M., Moeller, H., Kroetz, G., Berg, J.V. and Ziermann, R, *Sensors and Actuators*, **A74,** (1999), pp.56-9.
94. Wu, C.-H., Stefanescu, S., Kuo, H.-I., Zorman, C.A. and Mehregany, M., *Conf. Transducers '01*, vol.1, Munich, Germany, (2001), pp.514-7.
95. Pakula, L.S., Yang, H., Pham, H.T.M., French, P.J. and Sarro, P.M., *Conf. on Micro Electro Mechanical Systems*, Kyoto, Japan, (2003), pp.502-5.
96. Sarro, P.M., deBoer, C.R., Korkmaz, E. and Laros, J.M.W., *Sensors and Actuators,* **A67,** (1998), pp.175-80.
97. Okojie, R.S., Ned, A.A., Kurtz, A.D. and Carr, W.N., *Conf. International Electron Devices Meeting,* San Francisco, CA, USA, (1996), pp.525-8.

98. Okojie, R.S., Ned, A.A. and Kurtz, A.D., *Conf. Transducers '97,* vol.2, Chicago, IL, USA, (1997), pp.1407-9.
99. Okojie, R.S., Ned, A.A. and Kurtz, A.D., *Sensors and Actuators,* **A66,** (1998), pp.200-4.
100. Ned, A.A., Okojie, R.S. and Kurtz, A.D., *Conf. HITEC 1998*, Albuquerque, NM, USA, (1998), pp.257-60.
101. Shor, J.S., Okojie, R.S. and Kurtz, A.D., *Institute of Physics Conf. Series No 137*, Chapter 6, (1993), pp.523-6.
102. Okojie, R.S., Atwell, A.R., Kornegay, K.T., Roberson, S.L. and Beliveau, A., *Conf. MEMS 2002,* Las Vegas, NV, USA, (2002), pp.618-22.
103. Atwell, A.R., Okojie, R.S., Kornegay, K.T., Roberson, S.L. and Beliveau, A., *Sensors and Actuators,* **A104,** (2003), pp.11-18.
104. Figure(s) reprinted from ref. 103 with permission from *Elsevier.*
105. Beheim, G. and Salupo, C., *Materials Research Society Symp. Proc.*, **622,** (2000), pp.T8.9.1–16.
106. Katulka, G.L., *Conf.: IEEE SENSORS 2002*, vol.2, Orlando, FL, USA, (2002), pp.1134-8.
107. Nguyen, C.T.-C., Proc. IEEE Int. Symp. Circuits and Systems, (1997), pp.2825-8.
108. DeAnna, R.G., Roy, S., Zorman, C.A. and Mehregany, M., *I. Conf. on Modelling and Simulation of Microsystems, Semiconductors, Sensors and Actuators,* San Juan, Puerto Rico, (1999), pp.644-7.
109. DeAnna, R.G., Roy, S., Zorman, C.A. and Mehregany, M., *MST News,* no.5, (2000), pp.37-8.
110. Roy, S., DeAnna, R.G., Zorman, C.A. and Mehregany, M., *IEEE Trans. Electron Devices*, **49,** (2002), pp.2323-32.
111. Figure reprinted from ref. 110 with permission from *IEEE* (© 2002 IEEE)
112. Table reprinted from ref. 110 with permission from *IEEE* (© 2002 IEEE)
113. Kraus, T., Balzer, M. and Obermeier, E., *Conf. Transd. '97*, vol.1, Chicago, IL, USA, (1997), pp.67-70.
114. Yasseen, A.A., Wu, C.H., Zorman, C.A. and Mehregany, M, *IEEE Electron Device Letters*, **21,** (2000), pp.164-6.
115. Figure reprinted from ref. 114 with permission from *IEEE* (© 2000 IEEE)
116. Rajan, N., Zorman, C., Mehregany, M., DeAnna, R. and Harvey, R., *Conf. Proc. The Tenth Ann. Int. Workshop on Micro Electro Mechanical Systems,* Nagoya, Japan, (1997), pp.165-8.
117. Rajan, N., Zorman, C.A., Mehregany, M., DeAnna, R. and Harvey, R., *Thin Solid Films*, **315,** (1998), pp.170-8.
118. Rajan, N., Mehregany, M., Zorman, C.A., Stefanescu, S. and Kicher, T.P., *J. of Microelectromechanical Systems,* **8,** (1999), pp.251-7.
119. Figures reprinted from ref. 118 with permission from *IEEE* (© 1999 IEEE)

120. Gallis, S., Futschik, U., Castracane, J., Kaloyeros, AE., Efstathiadis, H., Sherwood, W., Hayes, S. and Fountzoulas, CG., Silicon Carbide 2002 – Mat. Proc and Dev., ed. Saddow S. E. *et al.* (MRS, Warrendale, PA, USA, 2002), pp.85-90.
121. Schmid, U., Eickhoff, M., Richter, C., Krotz, G. and Schmitt-Landsiedel, D., *Sensors and Actuators,* **A94,** (2001), pp.87-94.
122. Lundstrom, I., Shivaraman, M.S. and Svensson, C., *J. Appl. Phys.,* **46,** (1975), p.3876.
123. Neudeck, P.G., *J. Electronic Mater.,* **24,** (1995), p.283.
124. Savage, S. and Spetz, A.L.;, *Compound Semiconductor,* **6,** (2000), pp.76-81.
125. Spetz, A.L., Eriksson, M., Ekedahl, L.-G. and Lundstrm, I., *Conf. TATF 2000,* Nancy, France, (2000), pp.72-81.
126. Savage, S., Svenningstorp, H., Uneus, L., Kroutchinine, A., Tobias, P., Ekedahl, L.-G., Lundstrom, I., Harris, C. and Lloyd Spetz, A., *Materials Science Forum,* **353-356,** (2001), pp.747-52.
127. Hunter, G.W., Neudeck, P.G., Gray, M., Androjna, D., Chen, L.-Y., Hoffman, R.W. Jr., Liu, C.C. and Wu, Q., *Materials Science Forum,* **338-342,** (2000), pp.1439-42.
128. Chen, L.-Y., Hunter, G.W., Neudeck, P.G., Knight, D., Liu, C.C. and Wu, Q.H., *Conf. Ceramic Sensors III,* San Antonio, TX, USA, (1996), pp.92-106.
129. Spetz, A.L., Baranzahi, A., Tobias, P. and Lundström, I., *Phys. Stat. Sol (a),* **162,** (1997), pp.493-511.
130. Tobias, P., Rask, P., Göras, A., Lundström, I., Salomonsson, P. and Spetz, A.L., *Sensoren und Messysteme 2000,* VDI Berichte 1530, ISBN 3-18-091530-7, (VDI Verlag, Düsseldorf, 2000), pp.179-190.
131. Figure reprinted from ref. 130 with permission from *VDI Verlag*
132. Baranzahi, A., Tobias, P., Spetz, A.L., Lundström, I., Mårtensson, P., Glavmo, M., Göras, A., Nytomt, J., Salomonsson, P. and Larsson H., SAE Technical Paper Series 972940, *Combustion and Emission Formation in SI Engines,* SP-1300, (1998), pp. 231-240.
133. Figure reprinted from SAE Paper 972940 (ref. 132) with permission (© 1997/98 SAE International)
134. Svenningstorp, H., Widen, B., Salomonsson, P., Ekedahl, L.-G., Lundstrom, I., Tobias, P. and Spetz, A.L., *Sensors and Actuators,* **B77,** (2001), pp.177-85.
135. Hunter, G.W., Neudeck, P.G., Liu, C.C., Ward, B., Wu, Q.H., Dutta, P., Frank, M., Trimbol, J., Fulkerson, M., Patton, B. *et al., Conf.: IEEE SENSORS 2002,* vol.2, Orlando, FL, USA, (2002), pp.1126-33.
136. Uneus, L., Nakagomi, S., Linnarsson, M., Janson, M.S., Svensson, B.G., Yakimova, R., Syvajarvi, M., Henry, A., Janzen, E., Ekedahl, L.-G. *et al., Materials Science Forum,* **389-393,** (2002), pp.1419-22.
137. Linnarsson, M.K., Spetz, A.L., Janson, M.S., Ekedahl, L.G., Karlsson, S., Schöner, A., Lundström, I. and Svensson, B.G., *Materials Science Forum,* **338-342,** (2000), pp.937-40.

138. Nakagomi, S., Sanada, Y., Shinobu, H., Uneus, L., Lundstrom, I., Ekedahl, L.-G. and Spetz, A.L., *Conf. Transducers '01,* vol.2, Munich, Germany, (2001), pp.1758-61.
139. Zangooie, S., Arwin, H., Lundstrom, I. and Spetz, A.L., *Materials Science Forum,* **338-342,** (2000), pp.1085-8.
140. Mikalo, R.P., Hoffmann, P., Batchelor, D.R., Spetz, A.L., Lundstrom, I. and Schmeisser, D., *Materials Science Forum*, **353-356,** (2001), pp.219-22.
141. Nakagomi, S., Azuma, T. and Kokubun, Y., *Electrochemistry,* **70,** (2002), pp.174-7.
142. Nakagomi, S., Shinobu, H., Uneus, L., Lundstrom, I., Ekedahl, L.-G., Yakimova, R., Syvajarvi, M., Henry, A., Janzen, E. and Spetz, A.L., *Materials Science Forum*, **389-393,** (2002), pp.1423-6.
143. Svenningstorp, H., Tobias, P., Salomonsson, P., Lundström, I., Mårtensson, P. and Spetz A.L., *Sensors and Actuators*, **B57,** (1999), pp.159-165.
144. Lofdahl, M., Spetz, A.L., Eriksson, M. and Lundstrom, I., *Conf. TATF 2000,* Nancy, France, (2000), pp.382-4.
145. Nakagomi, S., Shindo, Y. and Kokubun, Y., *Physica Status Solidi,* **A185,** (2001), pp.33-8.
146. Jacobsen, S., Helmersson, U., Ekedahl, L.G., Lundstrom, I., Martensson, P. and Spetz, A.L., *Conf. Transducers '01*, vol.1, Munich, Germany, (2001), pp.832-5.
147. Chen, L.Y., Hunter, G.W., Neudeck, P.G. and Knight, D., *Solid State Electronics*, **42,** (1998), pp.2209-14.
148. Chen, L.Y., Hunter, G.W., Neudeck, P.G., Bansal, G., Petit, J.B. and Knight, D., *J. Vac. Sci. Technol.,* **A15,** (1997), pp.1228-34.
149. Nakagomi, S., Spetz, A.L., Lundstrom, I. and Tobias, P., *IEEE Sensors J.*, **2,** (2002), pp.379-86.
150. Hunter, G.W., Neudeck, P.G., Chen, L.Y., Knight, D., Liu, C.C. and Wu, Q.H., *NASA Technical Memorandum 107064*, (1995).
151. Schmeisser, D., Bohme, O., Yfantis, A., Heller, T., Batchelor, D.R., Lundstrom, I. and Spetz, A.L., *Physical Review Letters*, **83,** (1999), pp.380-3.
152. Savage, S.M., Konstantinov, A., Saroukhan, A.M. and Harris, C.I., *Materials Science Forum*, **338-342,** (2000), pp.1431-4.
153. Spetz, A.L., Tobias, P., Uneus, L., Svenningstorp, H., Ekedahl, L.-G. and Lundstrom, I., *Sensors and Actuators,* **B70,** (2000), pp.67-76.
154. Spetz, A.L., Uneus, L., Svenningstorp, H., Wingbrant, H., Harris, C.I., Salomonsson, P., Tengstrom, P., Martensson, P., Ljung, P., Mattsson, M. *et al.*, *Materials Science Forum*, **389-393,** (2002), pp.1415-18.
155. Svenningstorp, H., Uneus, L., Tobias, P., Lundstrom, I., Ekedahl, L.-G. and Spetz, A.L., *Materials Science Forum*, **338-342,** (2000), pp.1435-8.
156. Svenningstorp, H., Tobias, P., Salomonsson, P., Häggendal, B., Lundström, I., Ekedahl, L.-G. and Spetz, A.L., *Proc. Eurosensors XIV*, Copenhagen, Denmark, (2000), pp.933-936.

157. Wingbrant, H., Uneus, L., Andersson, M., Cerda, J., Savage, S., Svenningstorp, H., Salomonsson, P., Ljung, P., Mattsson, M., Visser, J.H. *et al.*, *Materials Science Forum*, **433-436,** (2003), pp.953-6.
158. Wingbrant, H., Svenningstorp, H., Salomonsson, P., Tengström, P., Lundström, I. and Spetz, A.L., *Sensors and Actuators,* **B93,** (2003), pp.295-303.
159. Cerda, J., Morante, J.R. and Spetz, A.L, *Mat. Science Forum*, **433-436,** (2003), pp.949-52.
160. Semancik, S., Cavicchi, R.E., Wheeler, M.C., Tiffany, J.E., Poirier, G.E., Walton, R.M., Suehle, J.S., Panchapakesan, B. and DeVoe, D.L., *Sensors and Actuators,* **B77,** (2001), pp.579-91.
161. Wiche, G., Berns, A., Steffes, H. and Obermeier, E., *Proc. Eurosensors 2004*, September 13-15, Rome, Italy, (2004)
162. Figures reprinted with permission from *TU Berlin, Microsensor and Actuator Technology.*
163. Solzbacher, F., Imawan, C., Steffes, H., Obermeier, E. and Eickhoff, M., *Sensors and Actuators,* **B77,** (2001), pp.111-115.
164. Solzbacher, F., Imawan, C., Steffes, H., Obermeier, E. and Eickhoff, M., *Sensors and Actuators,* **B78,** (2001), pp.216-220.
165. Figure reprinted from ref. 164 with permission from *Elsevier.*
166. Solzbacher, F., Imawan, C., Steffes, H. and Obermeier, E., *Proc. Eurosensors XIV*, Copenhagen, Denmark, (2000), pp.931-2.
167. Solzbacher, F., Imawan, C., Steffes, H., Obermeier, E. and Möller H., *Sensors and Actuators,* **B64,** (2000), pp.95-101.
168. Solzbacher, F., Imawan, C., Steffes, H. and Obermeier E., *Conf. Proc. Transducers '99*, Sendai, Japan, (1999), pp.1032-35.
169. Ho, J.J., Fang, Y.K., Wu, K.H., Hsieh, W.T., Chen, C.H., Chen, G.S., Ju, M.S., Lin, J.J. and Hwang, S.B., *Sensors and Actuators,* **B50,** (1998), pp.227-33.
170. Connolly, E.J., O'Halloran, G.M., Pham, H.T.M., Sarro, P.M. and French, P.J., *Sensors and Actuators,* **A99,** (2002), pp.25-30.
171. Mukai, T., Nagahama, S., Sano, M., Yanamoto, T., Morita, D., Mitani, T., Narukawa, Y., Yamamoto, S., Niki, I., Yamada, M., Sonobe, S., Shioji, S., Deguchi, K., Naitou, T., Tamaki, H., Murazaki, Y. and Kameshima, M., *Phys. Stat. Sol.,* **A200,** (2003), pp.52-7.
172. Dmitriev, V., Morozenko, Y., Popov, I., Suvorov, A., Syrkin, A. and Chelnokov, V., *Soviet Phys. - Tech. Phys., Lett.,* **12,** (1986), p.221.
173. Edmond, J., Kong, H., Suvorov, A., Waltz, D. and Carter, C. Jr., *Physica Status Solidi,* **A162,** (1997), pp.481-91.
174. Figures reprinted from ref. 173 with permission from *Wiley-VCH Verlag.*
175. Brown, D.M., Downey, E., Kretchmer, J., Mickon, G., Shu, E. and Schneider, D., *Solid-State Electronics*, **42,** (1998), pp.755-60.

176. Raffaelle, R.P., Bailey, S.G., Neudeck, P., Okojie, R., Schnabel, C.M., Tabib-Azar, M., Scheiman, D., Jenkins, P. and Hubbard, S., *Conf. 28th IEEE Photovoltaic Specialists Conf.*, Anchorage, AK, USA, (2000), pp.1257-60.
177. Hubbard, S.M., Tabib-Azar, M., Bailey, S., Rybicki, G., Neudeck, P. and Raffaelle, R., *Conf.: Twenty Sixth IEEE Photovoltaic Specialists Conf.,* Anaheim, CA, USA, (1997), pp.975-8.
178. Topic, M., Smole, F., Furlan, J. and Kusian, W., *J. Non-cryst. Solids,* **198-200,** (1996), pp.1180-1184
179. Cabrita, A., Pereira, L., Brida, D., Lopes, A., Marques, A., Ferreira, I., Fortunato, E. and Martins R., *Applied Surface Science*, **184,** (2001), pp.437-442.
180. Figure reprinted from ref. 179 with permission from *Elsevier.*
181. Cabrita, A., Figueiredo, J., Pereira, L., Águas, H., Silva, V., Brida, D., Ferreira, I., Fortunato E. and Martins R., *Applied Surface Science*, **184,** (2001), pp.443-447.
182. Vieira, M., Fernandes, M., Fantoni, A., Louro, P., Vygranenko, Y., Schwarz, R. and Schubert M., *Applied Surface Science,* **184,** (2001), pp.471-476.
183. Vlaskin, V.I., Berezhinsky, L.I., Vlaskina, C.I., Shin, D.H. and Kwon, K.H., *J. of the Korean Physical Society*, **42,** (2003), pp.391-3.
184. Kruangam, D., Toyama, T., Hattori, Y., Deguchi, M., Okamoto, H. and Hamakawa, Y., *Optoelect. Dev. and Tech.1*, **67,** (1986).
185. Gao, W., Lee, S.H., Xu, Y., Benson, D.K., Deb, S.K. and Branz, H.M., *Proc. of 2nd World Conf. on Photovoltaic Solar Energy Conversion,* Vienna, Austria, (1998).
186. Cho, P.S., Goldhar, J., Lee, C.H., Saddow, S.E. and Neudeck, P., *J. of Applied Physics*, **77,** (1995), pp.1591-9.
187. Saddow, S.E., Cho, P.S., Goldhar, J., Lee, C.H. and Neudeck, P.G., *Applied Physics Letters*, **65,** (1994), pp.3359-61.
188. Strokan, N.B., Ivanov, A.M., Savkina, N.S., Strelchuk, A.M., Lebedev, A.A., Syvajarvi, M. and Yakimova, R., *J. of Applied Physics*, **93,** (2003), pp.5714-9.
189. Richards, R.J. and De Los Santos, H.J., *Microwave J., Euro-Global Edition*, **44,** (2001), p.20, 24, 28, 32, 34, 38, 41.
190. De Los Santos, H.J. and Richards, R.J., *Microwave J., Euro-Global Edition*, **44,** (2001), p.142-4, 146, 148, 150, 152.
191. Melzak, J.M., *Conf. IEEE MTT-S International Microwave Symp. - IMS 2003*, vol.3, Philadelphia, PA, USA, (2003), pp.1629-32.
192. Huang, X.M., Ekinci, K.L., Yang, Y.T., Zorman, C.A., Mehregany, M. and Roukes, M.L., *Solid-State Sensor, Actuator and Microsyst. Workshop Hilton Head Island*, South Carolina, USA, (2002), pp.368-9.
193. Cleland, A.N. and Roukes, M.L., *Appl. Phys. Lett.*, **69,** (1996), pp.2653-5.
194. Network for Computational Nanotechnology, Research abstract on NEMS, http://ncn.purdue.edu/wps/portal/.cmd/cs/.ce/155/.s/1097/_s.155/1097.
195. Yang, Y.T., Ekinci, K.L., Huang, X.M.H., Schiavone, L.M., Roukes, M.L., Zorman, C.A. and Mehregany, M., *Applied Physics Letters*, **78,** (2001), pp.162-4.

196. Figure(s) reprinted from ref. 195 with permission from the *American Institute of Physics.*
197. Wicht Technology Consulting, Munich, Germany, 'The RF MEMS Market 2002 – 2007', http://www.wtc-consult.de/rfmems.pdf.
198. King J. A. (Ed.), *Materials Handbook for Hybrid Electronics*, (Artech House, Boston/London, 1988).
199. Sze S.M., *Physics of Semiconductor Devices*, (John Wiley and Sons, New York, 1981).
200. Heuberger A., *Mikromechanik*, (Springer, Berlin, 1991), p.58.
201. Landolt-Boernstein, New Series 3/22a: *Silicon Carbide*, (Springer, Heidelberg, 1985).

Appendix A: Material properties of silicon, 3C-SiC and 6H-SiC

Table 2. Material Properties of 3C-SiC and 6H-SiC in comparison to silicon.

Property	Silicon	3C-SiC	6H-SiC
Lattice Const. [Å]	5.43 (2)	4.359 (4)	a_0: 3.08 (2) c_0: 15.12 (2)
Bandgap [eV]	1.12 (2)	2.2 (4)	2.99 (2)
Density [g/cm^3]	2.33 (1)	3.21 (4)	3.21 (4)
Melting Point [°C]	1410 (1)	subl. at T>3100 (4)	subl. at T>3100 (4)
Knoop Hardness	1050 (1)	3300 (4)	2917 (4)
Hole Mobility [cm^2/Vs]	450 (2)	40 (4)	50 (2)
Electron Mob. [cm^2/Vs]	1500 (2)	1000 (4)	400 (2)
El. Satur. Drift Velocity [10^7 cm/s]	1 (2)	2.5 (4)	1.96 (4)
Breakdown Field [10^6 V/cm]	0.3 (2)	4 (4)	4 (4)
Thermal Conductivity [W/cm K]	1.5 (1)	4.9 (4)	4.9 (4)
Rel. Dielectric Constant	11.9 (2)	9.7 (4)	10 (2)
Lin. Thermal Expan. Coeff. [$1x10^{-6}$/K]	2.6 (1)	2.9 (4)	4.2 (4)
Elastic Moduli [10^{12} dyn cm^{-2}] c_{11} c_{12} c_{33} c_{44}	 1.65 (3) 0.63 (3) 0.79 (3)	 3.52 (4) 1.40 (4) 2.33 (4)	 5 (4) 0.92 (4) 5.64 (4) 1.68 (4)

References: (1) [198], (2) [199], (3) [200], (4) [201]